T0314493

Data Science with Semantic Technologies

Gone are the days when data was interlinked with related data by humans and human interpretation was required to find insights coherently. Data is no longer just data. It is now considered a Thing or Entity or Concept with meaning, so that a machine not only understands the concept but also extrapolates it the way humans do.

Data Science with Semantic Technologies: Deployment and Exploration, the second volume of a two-volume handbook set, provides a roadmap for the deployment of semantic technologies in the field of data science and enables the user to create intelligence through these technologies by exploring the opportunities and eradicating the challenges in the current and future time frame. In addition, this book offers the answer to various questions like: What makes a technology semantic as opposed to other approaches to data science? What is knowledge data science? How does knowledge data science relate to other fields? This book explores the optimal use of these technologies to provide the highest benefit to the user under one comprehensive source and title.

As there is no dedicated book available in the market on this topic at this time, this book becomes a unique resource for scholars, researchers, data scientists, professionals, and practitioners. This volume can serve as an important guide toward applications of data science with semantic technologies for the upcoming generation.

Data Science with Semantic Technologies

Deployment and Exploration

Edited by

Archana Patel and Narayan C. Debnath

CRC Press is an imprint of the
Taylor & Francis Group, an **informa** business

Designed cover image: Shutterstock

First edition published 2023
by CRC Press
6000 Broken Sound Parkway NW, Suite 300, Boca Raton, FL 33487–2742

and by CRC Press
4 Park Square, Milton Park, Abingdon, Oxon, OX14 4RN

CRC Press is an imprint of Taylor & Francis Group, LLC

Library of Congress Cataloging-in-Publication Data
Names: Patel, Archana (Lecturer in software engineering), editor. |
Debnath, N. C. (Narayan C.), editor.
Title: Data science with semantic technologies / edited by Archana Patel, Narayan C. Debnath.
Description: First edition. | Boca Raton : CRC Press, [2023] | Includes bibliographical references and index. | Contents: volume 1. New trends and future developments—volume 2. Deployment and exploration.
Identifiers: LCCN 2022056304 (print) | LCCN 2022056305 (ebook) | ISBN 9781032316666 (hardback ; volume 1) | ISBN 9781032316673 (paperback ; volume 1) | ISBN 9781032316680 (hardback ; volume 2) | ISBN 9781032316697 (paperback ; volume 2) | ISBN 9781003310785 (ebook ; volume 1) | ISBN 9781003310792 (ebook ; volume 2)
Subjects: LCSH: Big data. | Semantic computing. | Artificial intelligence. | Information science. | Information technology.
Classification: LCC QA76.9.B45 D397 2023 (print) | LCC QA76.9.B45 (ebook) | DDC 005.7—dc23/eng/20230109
LC record available at https://lccn.loc.gov/2022056304
LC ebook record available at https://lccn.loc.gov/2022056305

ISBN: 978-1-032-31668-0 (hbk)
ISBN: 978-1-032-31669-7 (pbk)
ISBN: 978-1-003-31079-2 (ebk)

DOI: 10.1201/9781003310792

Typeset in Times
by Apex CoVantage, LLC

Contents

Preface

Gone are the days when data was interlinked with related data by humans and human interpretation was required to find insights in a coherent manner. We are in an era where data is no longer just data. It is much more. Data of a domain of discourse is now considered a Thing or Entity or Concept with meaning so that the machine not only understand the concept but also extrapolates it the way humans do. Data science provides an invaluable resource that deals with vast volumes of data using modern tools and semantic techniques to find unseen patterns, derive meaningful information, and make business decisions. The roots of semantic technology and data science are as old as the modern digital computer. However, to create intelligence in data science, it becomes necessary to utilize the semantic technologies that allow machine-readable representation of data. This intelligence uniquely identifies and connects data with common business terms, and it also enables users to communicate with data.

Instead of structuring the data, semantic technologies help users to understand the meaning of the data by using the concepts of semantics, ontology, Web Ontology Language (OWL), linked-data, and knowledge-graphs. These technologies assist organizations in understanding of all the stored data, adding value to it, and enabling insights that were not available before. Organizations are also using semantic technologies to unearth precious nuggets of information from vast volumes of data and to enable more flexible use of data. These technologies can deal with data scientists' existing problems and help them to make better decisions for any organization. All these needs are focusing on a shift toward utilization of semantic technologies in data science that provide knowledge along with ability to understand, reason, plan, and learn with existing and new data sets. It also generates expected, reproducible user desired results.

This book aims to provide a roadmap for the deployment of semantic technologies in the field of data science and enable the user to create intelligence through these technologies by exploring the opportunities and eradicating the current and future challenges. In addition, this book intends to provide the answers to various questions like: What makes a technology semantic as opposed to other approaches to data science? What is knowledge data science? How does knowledge data science relate to other fields? This book explores the optimal use of these technologies to provide the highest benefit to the user under one comprehensive source and title. As there is no dedicated book available in the market on this topic at this time, this new book becomes a unique resource for scholars, researchers, data scientists, professionals and practitioners. Moreover, this book will provide a strong foundation to the other books which may be published in future on the same or closely related topic(s). This book, however, can have the advantage of early mover and will hold monopoly until a similar book is published.

Contributors

Y. Suresh Babu
Jagarlamudi Kuppuswamy Choudary College
Guntur, Andhra Pradesh, India

Ujwala Bharambe
Thadomal Shahani Engineering College
Mumbai, India

Ozgu Can
Department of Computer Engineering
Ege University
Izmir, Turkey

D.V. Chandrashekar
PG Department of Computer Science
Tellakula Jalayya Polisetty Somasundaram College
Guntur, India

Narayan C. Debnath
Department of Software Engineering
Eastern International University
Binh Duong Province, Vietnam

Sulochana Devi
Xavier Institute of Engineering
Mumbai, India

Marischa Elveny
Data Science & Computational Intelligence Research Group
Excellent Centre of Innovation and New Science Medan
Sumatera Utara, Indonesia

Syed Md Fazal
MANUU – Central University
Hyderabad, Telangana, India

Ing. Luis Ernesto Hurtado González
Universidad Central "Marta Abreu" de Las Villas (UCLV)
Santa Clara, Cuba

Johnbenetic Gnanaprakasam
Department of Computer Science
Thiruvalluvar University
Tamil Nadu, India

Garima Gujral
Centre for Library and Information Management Studies
Tata Institute of Social Sciences
Mumbai, Maharashtra, India

Elvina Herawati
Universitas Sumatera Utara
Sumatera Utara, Indonesia

Mokshitha Kanchamreddy
Vellore Institute of Technology
AP University
Amravati, India

Devi Kannan
Atria Institute of Technology
Bangalore, India

Farhana Kausar
Atria Institute of Technology
Bangalore, India

T.M. Kiran Kumar
Department of M.C.A
Siddaganga Institute of Technology
Tumkur, Karnataka, India

Ravi Lourdusamy
Department of Computer Science
Thiruvalluvar University
Tamil Nadu, India

C. Amed Abel Leiva Mederos
Universidad Central "Marta Abreu" de Las Villas (UCLV)
Santa Clara, Cuba

Musa Milli
Computer Engineering Department
Turkish Naval Academy
National Defense University
Istanbul, Turkey

Chhaya Narvekar
Xavier Institute of Engineering
Mumbai, India

Mahyuddin K.M. Nasution
Data Science & Computational Intelligence Research Group
Excellent Centre of Innovation and New Science Medan
Sumatera Utara, Indonesia

Karthika Natarajan
Vellore Institute of Technology
AP University
Amravati, India

Aishwarya P.
Atria Institute of Technology
Bangalore, India

Archana Patel
National Forensic Sciences University
Gandhinagar, Gujarat, India

Suryakant Sawant
Tata Consultancy Services
Mumbai, India

Fatmana Şentürk
Computer Engineering Department
Pamukkale University
Denizli, Turkey

J. Shivarama
Centre for Library and Information Management Studies
Tata Institute of Social Sciences
Mumbai, Maharashtra, India

K. Suneetha
PG Department of Computer Science
Tellakula Jalayya Polisetty Somasundaram College
Guntur, India

Mamatha T.
Atria Institute of Technology
Bangalore, India

1 Machine Learning Meets the Semantic Web

D.V. Chandrashekar, Y. Suresh Babu, K. Suneetha and T.M. Kiran Kumar

CONTENTS

1.1 INTRODUCTION

Two major developments in the history of the World Wide Web meet at the intersection of semantic web services (SWS). One is the quick evolution of web services and the second is the semantic web, both of which are examples of technologies. Greater emphasis is placed in semantic web on the dissemination of more semantically expressive metadata in a collaborative knowledge structure, which allows for the distribution of software agents that wisely exploit the Internet's offerings. Motivation for web services rely on having vendor-neutral software that can communicate with other systems that may be very different from one another. Using infrastructure as a platform of network with various levels to obtain the goal. The other primary goal is the driving force behind the expansion of web services capability to coordinate business processes involving deployment of heterogeneous parts (as services) throughout property lines. As a result of these goals, construction of generally accepted web service standards, in addition to Business Process Execution Language (BPEL); Web Services Description Language (WSDL); and Universal Description, Discovery, and Integration (UDDI) with the help of semantic web services, automated.

Web service discovery at the edge using rich semantic models. Using this model will, despite the facts included in SWS, allow automatic resource allocation for defining the project clearly, it is also used in finding the independency and projection of SWS in a specified environment. Among the most important issues with the current framework include the fact that UDDI does not capture the interconnections of its

DOI: 10.1201/9781003310792-1

database of entities, hence it is unable to use the search-related inferences based on semantic information.

In machine learning, a function can be described in one of two main ways: supervised learning or unsupervised learning. The variables in supervised learning can be categorized as explanatory variables and a single or multiple dependent variables. Similar to regression analysis, the goal of this analysis is to establish a causal connection between independent and dependent variables. To use directed data mining methods, it is necessary to already have a good idea of the values of the dependent variable across a substantial portion of the dataset. In unsupervised learning, both independent and dependent variables are given equal weight. In order for supervised learning to work, the target variable must be clearly specified, and a sufficient sample of its possible values must be provided. In most cases of unsupervised learning, either the target variable is unobservable or there is insufficient data to make any meaningful predictions.

KGs are becoming increasingly popular as a means of storing vast amounts of data on the web, where they serve a key role in applications like search engines. Google began using KGs as a search tactic in 2012 when they were initially introduced. Word processing has taken on a symbolic form thanks to these graphs. Semantic web (SW) developers are particularly interested in KGs because they are employed in social networking and e-commerce applications alike.[1] In spite of the fact that there is no single KG definition, it is possible to define it as a method of portraying real-world entities and occurrences in addition to conceptual ideas in their abstract forms. Even though KGs are simple to understand and contain a wealth of knowledge about the world, they are difficult to employ in ML. For semantic service discovery to take place, however, we only need a small amount of metadata to be made available that describes the web services' functions and can be used by machine learning methods for tasks like classification, clustering, and association mining. In this chapter, we contribute by providing a comprehensive overview of SWS discovery frameworks based on machine learning approaches, methodologies, and techniques applied to the semantic discovery of web services, and we discuss the limitations of these methods and the ways they might be improved in the future.

The subfields that fall under the umbrella term "artificial intelligence." In order for these systems to be able to perform their duties more effectively, it is necessary to construct ones that are capable of learning from data samples taken from the past as well as from the real world.[2] The combination of ML and KG moves quickly. On the one hand, machine learning approaches boost various data-driven tasks' accuracy to a remarkable degree. On the other hand, KGs can express knowledge about entities and their interactions in a way that is both highly reliable and understandable. The combination of knowledge graphs and ML is expected to make systems more accurate, easier to understand, and reusable.[3] The following is a breakdown of the chapter's structure: Section 1.2 introduces KGs, whereas Section 1.3 focuses on the relationship between KGs and ontologies. A link is established between KGs and ML, specifically in the fourth paragraph. Section 1.5 examines the connection between KGs and deep learning, while Section 1.6 explores the domain's unresolved concerns and challenges. Section 1.7 is the final section of the chapter.

1.2 KNOWLEDGE GRAPHS

There have been numerous attempts, with varied degrees of success, to describe what knowledge graph. An unavoidable blending of discrepancies has occurred as a result of various definitions in the literature. The Wikipedia definition is just one of many relevant and more recent ones[4–6] that have been presented by many researchers. One way of organising and visualising the data in a knowledge base (KB) is to use a multidomain graph in which the nodes reflect the many entities of interest. Figure 1.1 shows one example.

An example of a knowledge graph is shown in Figure 1.1. The nodes are the entities, the edge labels are the types of relationships, and the edges are the relationships that already exist.

By using a reasoner and ontology, according to Ehrlinger and Wöß (2016), KGs can acquire new knowledge from existing data.[7] This definition assumes that an advanced knowledge generator (KG) is better than a basic knowledge generator (BG) because it generates new knowledge via a reasoning engine and supports a wide range of information sources (KB). A "large" graph that is not defined by this definition is left out. Farber et al. (2018) and Huang et al. (2017) have shown that the Resource Description Framework (RDF) is used to describe KGs (RDF).[8-9] Paulheim (2017) claims that KGs cover a wide range of topics and are not restricted to a specific classification.[10] Additional research support enhance how knowledge is stored and analysed using KGs. The KGs can be used to break it down into more well-known web-based data publication strategies such as linked data. Other well-known KGs that are hidden from public view include those operated by Google, Microsoft, and the social media behemoth Facebook. DBpedia and Wikidata KGs are two additional well-known and frequently utilised KGs that are available to the general public. There are several choices available.[11]

Most KGs are unable to obtain precise references for the extraction methods they employ, as well as the overall design, visualisation, and storage of all this

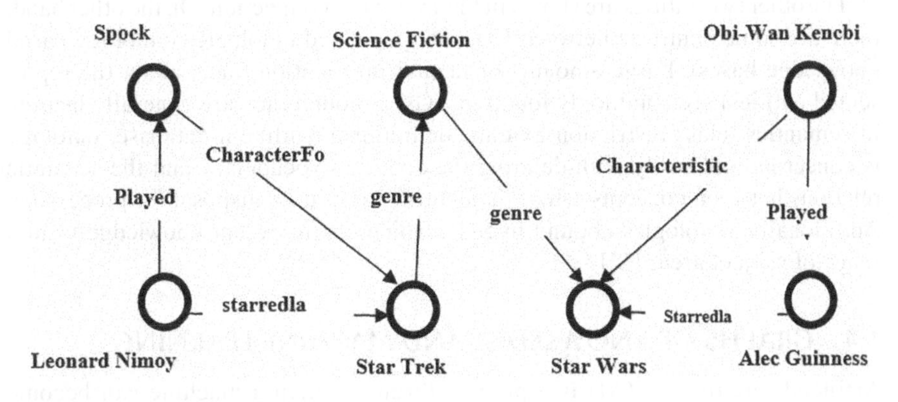

FIGURE 1.1 Knowledge graph.

knowledge.[12] Creating a KG can be accomplished in one of two ways: top-down (schema-driven) or bottom-up (data-driven). After creating ontology and a schema, data can then be put into the graph. In the second step, information is gathered from a variety of sources (such as text documents, databases, and open data links) and then combined to form the KG schema.

Facebook KG is a crucial tool that allows linked people to use internal search within the Facebook network to obtain increasingly accurate results. KG, or Google's Knowledge Graph, is a useful feature that pulls data from a variety of sources, including Wikipedia, in order to provide users with more relevant search results. An enormous knowledge base has been generated by analysing Wikipedia articles to make them available to everyone on the World Wide Web. Due to the sheer number of pages on Wikipedia, as well as those written in different languages, there are a number of inconsistencies in the material. As a result of Wiki-data KG, the challenge of handling this data was finally resolved, allowing information to be connected across several languages at the same time. Confusion can be tolerated because of a system that appropriately organizes all the information.

1.3 GRAPHS OF KNOWLEDGE AND ONTOLOGY

Domain ontologies are formal and clear descriptions of how people agree to think about a subject. Ontologies have been developed specifically for museums, for security, and for surveillance, just to name a few. Ontology can provide schemas and real data, or it can describe schemas and real data (classes, their relationships, and class restriction axioms). Alternatively, a populated ontology is one in which the classes are already filled in. On the other hand, a knowledge base for domain-specific entities is referred to as ontology since the knowledge it contains pertains to instances or individuals of its ontological classes. There is a lot of overlap between KGs and populated ontologies. Both use the Resource Description Framework (RDF) to describe their data. Both use semantic relations (links or edges) between things to convey domain knowledge (nodes). Subject, predicate, and object (SPO) statements are used to convey knowledge about certain things.

The other two entities are shown in Figure 1.2 in a connection. On the other hand, there are some contrasts between KGs and populated ontologies (ontology-based knowledge bases). Large amounts of factual information (facts about the represented entities) are commonly found in KGs, although they are generally lacking in semantics (class restriction axioms, definitions). Unlike a database, ontology is concerned primarily with defining the domain's vocabulary and the semantic relations between concepts rather than putting data at its disposal for processing. Additionally, ontologies abound in KGs, which can represent knowledge from a range of subject areas.[13]

1.4 GRAPHS OF KNOWLEDGE AND MACHINE LEARNING

Artificial intelligence (AI) is a process through which a machine can become intelligent by learning from the information that is fed into the process of data analysis. The term "ML method" refers to a set of algorithms that can be used to

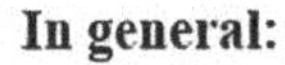

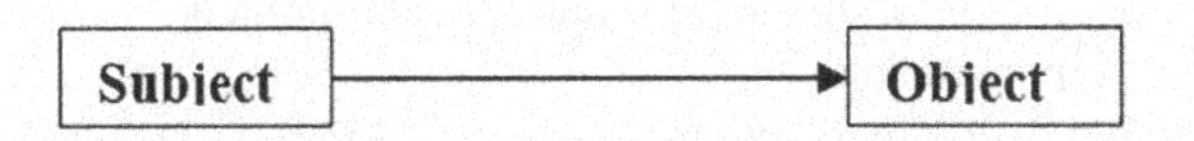

Example : < Lecnard Nimov , starrredIn, Star Trek

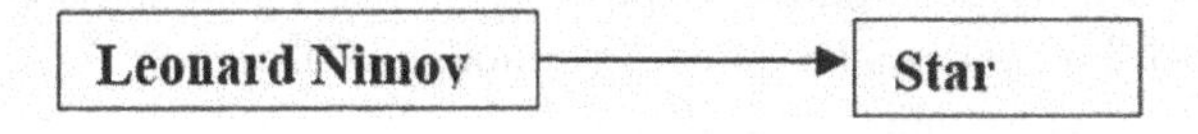

FIGURE 1.2 Three times the accusation that Leonard Nimoy appeared in the *Star Trek* film is made.

solve a certain computational problem. Using this approach, you can deal with any limitations that the problem may have. The three most common types of machine learning are supervised learning, semi-supervised learning, and reinforcement learning techniques. The process of learning is broken up into three steps: collecting data, processing and analysing the data, and using the results of the processing.

ML approaches in KGs[2] can be used to handle problems such as the prediction of types and connections, the enrichment and integration of ontologies, and so on. In particular, machine learning strategies geared at the SW were developed as a solution to this issue. With reference to failure of reasoning of abstract based on the content of ontology in reference to domain and erroneous content. These massive KGs, or "knowledge-based graphs," store actual data in the form of connections between different things. Triples are mechanically extracted from unstructured text in an automated knowledge base creation process.

These are the methods which are used in NLP and machine learning process. Because of their reliance on human expertise, all other methods have either restrictions or are not scalable. In recent years, more emphasis has been placed on automated methods. In order to make the web more valuable to machines,[14] semantic web technology aims to add metadata to websites. The OWL format and the ontology's reasoning abilities are used to keep track of this metadata. On the other hand, metadata management has a variety of issues. Inconsistent and noisy data, lack of standards, and the time it takes to develop ontology are just a few examples.

As a result of this, ontology refinement and enrichment concerns have also surfaced. In ML, the first three are classified as classification issues, while the fourth and final one is categorised as a challenge of concept learning. Deep learning is needed in order to address the semantic web's classification and concept learning issues. Techniques based on numbers have been shown to be useful through embeddings.[15] Symbolic and numeric-based ML approaches are the two most common types of machine learning (ML) techniques. Methods that approach the semantic web challenge from a logical standpoint fall under the symbol-based category.

For example,

I. retrieval and concept acquisition for ontology enrichment,
II. knowledge completion and
III. the learning of disjointness axioms are included in this category.

If an individual is an instance of a particular idea, then the instance retrieval problem has been addressed by a classification problem. When KNN and SVM were first introduced,[16–18] techniques based on Terminological Decision Trees (TDT) were made that were easier to understand.[19, 20]

Enrichment of ontology and learning concept descriptors are the primary goals of concept learning for this challenge. Supervised idea learning is used to deal with this issue, which resembles an intentional problem. For logical representations, in this category, there are a number of approaches.[21–25] The purpose of the knowledge completion problem is to locate information that is not already in the knowledge base. AMIE[26, 27] is an example of a suggestive method for dealing with this issue. AMIE intends to scan RDF knowledge repositories for logic rules to better anticipate future statements. According to a study that was recently published,[27] the approach attempts to derive SWRL rules using OWL ontologies. This strategy aims to uncover the axioms from the data that have been neglected throughout the modelling process, which can lead to a misinterpretation of the target domain's negativity. Other authors[28, 29] have offered suggestions for how to approach this challenge in an instructive manner. The correlation co-efficient, negative and association rules, and correlation rules are all studied using these techniques.

Methods designed to link a prediction problem to numerical data fall under the numeric-based category. Classification is used to determine whether or not a set of triplets exists, and the results are often presented in RDF format. Up until this point in time, the probabilistic latent models and embedding models have been presented as possible solutions to this issue. The Infinite Hidden Semantic Model (IHSM) is a suggestive approach for a probabilistic latent model.[30] Latent class variables are associated with each resource/node in this model, and first-order in the learning process logic constraints are used. Like probabilistic models, these models use an embedding vector to keep track of the hidden properties of a resource as data is collected. As an example of an embedding model, RES-CAL has been proposed.[31] The multi-graph structure of the data set can be seen by using this method, which involves the factorization of an adjacency tensor in three different ways

1.5 THE COMBINATION OF KNOWLEDGE GRAPHS AND DEEP LEARNING

Deep learning was first introduced a few years ago. There is a wide range of tasks involving machine learning, from the classification of visual scenes and video analysis to the interpretation and identification of natural speech and language. Images, speech signals, feature vectors, and other data types commonly employed. All of these endeavours are represented in Euclidean space. Convolutional neural networks (CNNs), recurrent neural networks (RNNs), long-short-term memory (LSTM), and auto-encoders are all deep learning techniques that work well with the data types listed above. The data-driven shapes of much deep learning-based architecture are projected into a space that is not Euclidean, especially as KGs.[32–34] The inter-action assumption is broken by this data structure, which uses a linking strategy. As a result, each node gets linked to a variety of other nodes and variant kinds throughout the linking process between things and their attributes.

KG nodes may have different neighbourhood sizes, and the interactions between them may be different. As well, there is a wide range of options. To deal with the aforementioned complexity, new neural network topologies referred to as "Graph Neural Networks"[35] have been developed.

Figure 1.3 shows the two convolutional layers of Conventional Geometry Neural Network (ConvGNN). Within each convolutional layer, the hidden representation of a node is encapsulated by gathering feature information from the nodes that are neighbouring it. When the outputs are ready, they are activated using rectified linear unit (ReLu).[36] Each node's final hidden representation receives messages from a neighbouring community.[37] After their introduction in 2005, Scarselli et al. (2008) [38] and Gallicchio and Micheli (2010)[39] have evolved graph neural networks further. Gori et al. (2005),[40] have repeatedly propagated input to one or more neighbouring nodes, and graph neural networks (GNNs) train the representation of a target node until a stable fixed point is discovered. There has recently been an increase in efforts to overcome this limitation.[41, 42] This research focuses on recurrent neural graph networks (RecGNNs). Because of the extensive use of convolutional neural networks in computer vision, there has been a proliferation of approaches that use a technique called "convolution" to analyse the graph data. Convolutional graphs are a subfield that falls under the umbrella of graph neural networks (ConvGNNs). Figure 1.3 is a representation of a ConvGNNs design that serves as an example. Convolutional graph neural network-based spectral and spatial techniques are often categorised into two key subcategories. There has been some discussion over the use of indicative spectral-based categorization, as advocated by Bruna et al. [43] In subsequent studies,[43–46] convolutional graph neural networks were improved and expanded.

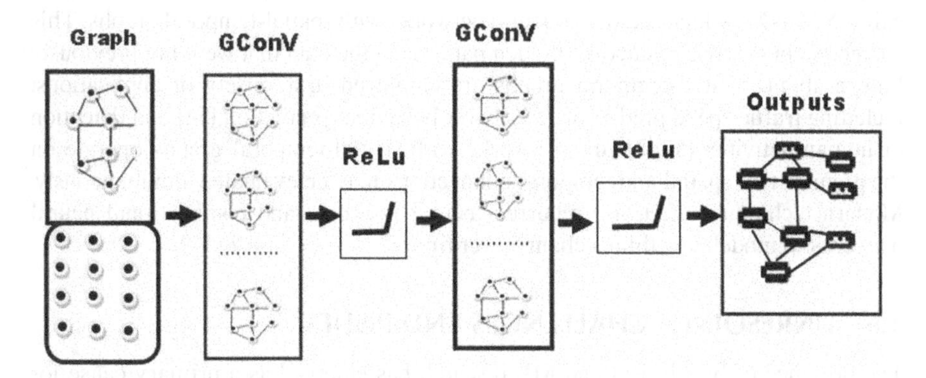

FIGURE 1.3 Two way convolutional layers.

More work has been done on spectral-based graph neural networks than on spatial-based ones. Micheli [47] tackled the problem of graph mutual dependence for the first time in 2009. They did this by first incorporating forwarding techniques from recurrent graph neural networks into composite non-recursive layers. Convolutional graph neural networks, also known as CGNNs, have been constructed and put into use in recent years.[48–50] Alternative graph neural networks, such as graph autoencoders (GAEs) and spatial-temporal graph neural networks (STGNs), have emerged in the last few years (STGNNs). These frameworks for graph modelling can be made with neural networks like recurrent and convolutional networks. Wu and his colleagues (2020)[37] published a taxonomy of graph neural networks as part of their substantial research on graph neural networks. There are four main types of graph neural network architecture. The initial class of graph neural networks is represented by Rec-GNNs. Recurrent graph neural networks use neural topologies with recurrent connections to learn node representations. Nodes in networks should communicate continually until a stable equilibrium is attained, as this is what is expected of them in theory. Convolutional graph neural networks were motivated by the recurrent graph neural network concept. Spatial-Based Convolutional Graph Neural Networks, in particular, are derived from the idea of a message being transmitted via space.

Using convolutional graph neural networks, convolution can also be extended from grids to graphs (aka ConvGNNs). In order to create a node's representation, it is necessary to take into account all of the features of the node and its neighbours. Convolutional graph neural networks use numerous layers of convolution to build high-level representations of nodes. When developing more complex models, convolutional graph neural networks proved to be an excellent starting point. The third group includes the GAEs, which are trained using unsupervised learning in the absence of labelled input. They also encode the graph data into a latent vector space before decoding it and recreating the original graph from it. Graph autoencoders may be a better technique to learn network embeddings and graph generative distributions (GAEs).[51, 52]

Graph embedding can be accomplished by utilising a graph autoencoder to produce information regarding the structure of the graph, such as the graph adjacency matrix. Either individual nodes or edges of a graph can be added one at a time, or the entire network can be constructed all at once. The final type of neural network is called STGNNs, which stands for neural networks with spatial-temporal graphs. This category's purpose is to uncover hidden patterns in the data that were not previously known about. Spatial-temporal graphs are employed in a variety of applications, including traffic speed prediction,[53] driver behaviour prediction,[54] and detection of human activities.[55] Neural networks with spatial-temporal graphs operate on the premise that spatial and temporal connections may be evaluated simultaneously. Modern techniques often use recurrent neural networks and convolutional neural networks to model how things change over time.

1.6 UNRESOLVED CHALLENGES AND ISSUES

The presence of bias in KG and ML research has emerged as a primary cause for concern and obstruction in recent years. A common example of study bias is the effect of preconceptions on the results of research, especially in the field of artificial intelligence. Although the ML method and the algorithm that employs them are prone to prejudice. Despite the fact that data and algorithms are created by humans, they are susceptible to human error and bias. Data samples employed in algorithmic analysis are not reliable representations of their respective datasets when data is subjective or biased. Using Google's image search for "CEO," this information may be utilised to train an intelligent system to suggest that doctors and nurses are respectively good jobs for men and women, which recently displayed an example of representational bias. While the ML/DL community has been working to minimise representational biases, the KG and semantic web communities have not yet taken notice. In the current cloud of linked open data, sample bias may not be an issue (LOD). Open and commercial KGs, on the other hand, receive distorted data. As ML-based techniques become more common in AI systems and applications, it will become harder to remove bias from KGs (data and schema). Data and schema biases must be taken into account when attempting to remove KGs of prejudice. DBpedia's KG, for example, does not include all of the accessible data about the world's entities, whether spatial or non-spatial (all the world as we know it). Instead, Europe and the United States are clearly covered by more data than Asia. The schema/ontology level, on the other hand, is highly subject to bias because most ontology is built using a top-down methodology, typically with application requirements in mind. This is when knowledge engineers and domain experts work together to spread design patterns based on their subjective engineering decisions (human-centered approach). Combining bottom-up (data-driven) ontological engineering with machine learning algorithms that can extract and learn on the fly is another way to use this technique. As previously explained, bias remains, as it is likely amplified by skewed data. We apply logical axioms or rules for engineering the datasets using sample data. So it's important to come up with a new way to build KGs that don't have any bias, along with tools for managing them that are in line with current policies and standards for reducing AI bias. This technique (detailed phases, processes, and activities) is likely to lead to more unbiased AI apps. In an age where AI applications are heavily focused on ML/DL and KGs, debiasing them must be a top priority and a constant worry.

1.7 CONCLUSION

This chapter explored a number of relevant applications for KGs and ML/DL, and the relationship between the two has been explained. The interplay between ontology and KGs was also highlighted in this chapter. To further comprehend the aforementioned correlations, a full presentation of methods for machine learning that are based on symbols as well as numbers has been demonstrated. Deep learning and neural networks are the foundations of graph neural networks, a strong and useful tool for machine learning applications in the graph domain. These two methodologies are combined by KGs to produce graph neural networks. Recurrent graph neural networks were classified into four types in this chapter: recurrent graph neural networks, convolution graph neural networks, graph auto-encoders, and spatial-temporal graph neural networks. The chapter then examined unresolved concerns and problems within the scope of this research. It is critical to emphasise the impact of KG bias at the schema, ontology, and data levels. People have said that ML/DL-based AI systems will only work if the debasing of KGs is given enough attention.

REFERENCES

1. Bonatti, P. A., Decker, S., Polleres, A., & Presutti, V. (2019). Knowledge graphs: New directions for knowledge representation on the semantic web (dagstuhl seminar 18371). *Dagstuhl Reports*, 8(9). Schloss Dagstuhl-Leibniz-Zentrum fuer Informatik.
2. d'Amato, C. (2020). Machine learning for the semantic web: Lessons learnt and next research directions. *Semantic Web*, no. Preprint, 1–9.
3. Nickel, M., Murphy, K., Tresp, V., & Gabrilovich, E. (2015). A review of relational machine learning for knowledge graphs. *Proceedings of the IEEE*, 104(1), 11–33.
4. Marchi, E. and Miguel, O. (1974). On the structure of the teaching-learning interactive process. *International Journal of Game Theory*, 3(2), 83–99.
5. van den Berg, H. (1993). First-order logic in knowledge graphs. *Current Issues in Mathematical Linguistics*, 56, 319–328.
6. Bakker, R. R. (1987). *Knowledge graphs: Representation and structuring of scientific knowledge*. IGI publishing house, Hershey, Pennsylvania, USA.
7. Ehrlinger, L., & Wöß, W. (2016). Towards a definition of knowledge graphs. *SEMANTiCS (Posters, Demos, SuCCESS)*, 48, 1–4.
8. Farber, M., Bartscherer, F., Menne, C., & Retting-er, A. (2018). Linked data quality of dbpedia, freebase, open-cyc, wikidata, and yago. *Semantic Web*, 9(1), 77–129.
9. Huang, Z., Yang, J., van Harmelen, F., & Hu, Q. (2017). Constructing diseasecentric knowledge graphs: A case study for depression (short version). In *Conference on artificial intelligence in medicine in Europe*. Springer Nature, Italy, pp. 48–52.
10. Paulheim, H. (2017). Knowledge graph refinement: A sur-vey of approaches and evaluation methods. *Semantic Web*, 8(3), 489–508.
11. Bizer, C., Heath, T., & Berners-Lee, T. (2011). Linked data: The story so far. In *Semantic services, interoperability and web applications: Emerging concepts*. IGI Publishing House, Academic Publications Ltd., Bulgaria, pp. 205–227.
12. Zhao, Z., Han, S.-K., & So, I.-M. (2018). Architecture of knowledge graph construction techniques. *International Journal of Pure and Applied Mathematics*, 118(19), 1869–1883.
13. McCusker, J. P., Erickson, J., Chastain, K., Rashid, S., Weerawarana, R., & McGuinness, D. (2018). What is a knowledge graph. *Semantic Web Journal*, 2–14.
14. d'Amato, C., Fanizzi, N., & Esposito, F. (2010). Inductive learning for the semantic web: What does it buy? *Semantic Web*, 1(1, 2), 53–59.
15. Deng, L., & Yu, D. (2014). Deep learning: Methods and applications. *Foundations and Trends in Signal Processing*, 7(3–4), 197–387.
16. d'Amato, C., Fanizzi, N., & Esposito, F. (2008). Query answering and ontology population: An inductive approach. In *European semantic web conference*, Springer Nature, Spain, pp. 288–302.
17. Rettinger, N. A., Losch, U., Tresp, V., d'Amato, C., & Fanizzi, N. (2012). Mining¨ the semantic web. *Data Mining and Knowledge Discovery*, 24(3), 613–662.
18. Bloehdorn, S., & Sure, Y. (2007). Kernel methods for mining instance data in ontologies. In *The semantic web*. Malaysia: Springer Nature, pp. 58–71.
19. Fanizzi, N., d'Amato, C., & Esposito, F. (2010). Induction of concepts in web ontologies through terminological decision trees. In *Joint European conference on machine learning and knowledge discovery in databases*. Springer Nature, Singapore, pp. 442–457.

20. Rizzo, G., Fanizzi, N., d'Amato, C., & Esposito, F. (2018). Approximate classification with web ontologies through evidential terminological trees and forests. *International Journal of Approximate Reasoning*, 92, 340–362.
21. Fanizzi, N., d'Amato, C., & Esposito, F. (2008). Dl-foil concept learning in description logics. In *International conference on inductive logic programming*. Springer LNCS Publications, Germany, pp. 107–121.
22. Tran, A. C., Dietrich, J., Guesgen, H. W., & Marsland, S. (2012). An approach to parallel class expression learning. In *International workshop on rules and rule markup languages for the semantic web*, Vol. 2. Springer, Berlin Heidelberg, Germany, pp. 302–316.
23. Lehmann, J., Auer, S., Buhmann, L., & Tramp, S. (2011). Class expression learning for ontology engineering. *Journal of Web Semantics*, 9(1), 71–81.
24. Rizzo, G., Fanizzi, N., d'Amato, C., & Esposito, F. (2018). A framework for tackling myopia in concept learning on the web of data. In *European knowledge acquisition workshop*. Springer, Berlin Heidelberg, Germany, pp. 338–354.
25. Tran, A. C., Dietrich, J., Guesgen, H. W., & Marsland, S. (2017). Parallel symmetric class expression learning. *The Journal of Machine Learning Re-search*, 18(1), 2145–2178.
26. Baader, F., Calvanese, D., McGuinness, D., Pa-tel-Schneider, P., Nardi, D., et al. (2003). *The description logic handbook: Theory, implementation and applications*. Cambridge: Cambridge University Press.
27. d'Amato, C., Tettamanzi, A. G., & Minh, T. D. (2016). Evolutionary discovery of multi-relational association rules from ontological knowledge bases. In *European knowledge acquisition workshop*. Springer Nature, Berlin/Heidelberg, Germany, pp. 113–128.
28. Volker, J., Fleischhacker, D., & Stuckenschmidt, H. (2015). Automatic acquisition of class disjointness. *Journal of Web Semantics*, 35, 124–139.
29. Volker, J., & Niepert, M. (2011). Statistical schema induction. In *Extended semantic web conference*. Springer LNCS Publication, Berlin/Heidelberg, Germany, pp. 124–138.
30. Rettinger, A., Nickles, M., & Tresp, V. (2009). Statistical relational learning with formal ontologies. In *Joint European conference on machine learning and knowledge discovery in databases*. Springer, pp. 286–301.
31. Nickel, M., Tresp, V., & Kriegel, H.-P. (2011). A three-way model for collective learning on multi-relational data. In *Icml*. Proceedings of the 28th International Conference on Machine Learning, Bellevue, WA, USA, 2011 Canada: HSG Publication.
32. LeCun, Y., Bengio, Y. et al. (1995). Convolutional networks for images, speech, and time series. In *The handbook of brain theory and neural networks*, vol. 3361, no. 10, p. 1995.
33. Schmidhuber, J., & Hochreiter, S. (1997). Long short-term memory. *Neural Computation*, 9(8), 1735–1780.
34. Hochreiter, S., & Schmidhuber, J. (1997). Long short-term memory. *Neural Computation*, 9(8), 1735–1780.
35. Vincent, P., Larochelle, H., Lajoie, I., Bengio, Y., Manzagol, P.-A., & Bottou, L. (2010). Stacked denoising auto-encoders: Learning useful representations in a deep network with a local denoising criterion. *Journal of Machine Learning Research*, 11(12).
36. Gao, Y., Li, Y.-F., Lin, Y., Gao, H., & Khan, L. (2020). Deep learning on knowledge graph for recommender system: A survey. arXiv preprint arXiv:2004.00387.
37. Wu, Z., Pan, S., Chen, F., Long, G., Zhang, C., & Philip, S. Y. (2020). A comprehensive survey on graph neural networks. *IEEE Transactions on Neural Networks and Learning Systems*, 4, 30–39.
38. Scarselli, F., Gori, M., Tsoi, A. C., Hagenbuchner, M., & Monfardini, G. (2008). The graph neural network model. *IEEE Transactions on Neural Networks*, 20(1), 61–80.
39. Gallicchio, C., & Micheli, A. (2010). Graph echo state networks. In *The 2010 international joint conference on neural networks (IJCNN)*. IEEE, Manhattan, New York, U.S. pp. 1–8.
40. Gori, M., Monfardini, G., & Scarselli, F. (2005). A new model for learning in graph domains. In *Proceedings. 2005 IEEE international joint conference on neural networks*, vol. 2. IEEE Explorer, Germany, pp. 729–734.

41. Li, Y., Tarlow, D., Brockschmidt, M., & Zemel, R. (2015). Gated graph sequence neural networks. arXiv pre-print arXiv:1511.05493.
42. Dai, H., Kozareva, Z., Dai, B., Smola, A., & Song, L. (2018). Learning steady states of iterative algorithms over graphs. In *International conference on machine learning*. Proceedings of the 35th International Conference on Machine Learning, Stockholm, Sweden, pp. 1106–1114.
43. Bruna, J., Zaremba, W., Szlam, A., & LeCun, Y. (2013). Spectral networks and locally connected networks on graphs. arXiv preprint arXiv:1312.6203.
44. Henaff, M., Bruna, J., & LeCun, Y. (2015). Deep convolutional networks on graph-structured data. arXiv preprint arXiv:1506.05163.
45. Defferrard, M., Bresson, X., & Vandergheynst, P. (2016). Convolutional neural networks on graphs with fast localized spectral filtering. arXiv preprint arX-iv:1606.09375.
46. Kipf, T. N., & Welling, M. (2016). Semi-supervised classification with graph convolutional networks. arXiv preprint arXiv:1609.02907.
47. Micheli, A. (2019). Neural network for graphs: A contextual constructive approach. *IEEE Transactions on Neural Networks*, 20(3), 498–511.
48. Levie, R., Monti, F., Bresson, X., & Bronstein, M. M. (2018). Cayleynets: Graph convolutional neural networks with complex rational spectral filters. *IEEE Transactions on Signal Processing*, 67(1), 97–109.
49. Agarap, A. F. (2018). Deep learning using rectified linear units (relu). arXiv preprint arXiv:1803.08375.
50. Atwood, J., & Towsley, D. (2015). Diffusion-convolutional neural networks. arXiv preprint arXiv:1511.02136.
51. Niepert, M., Ahmed, M., & Kutzkov, K. (2016). Learning convolutional neural networks for graphs. In *International conference on machine learning*. Proceedings of the 33rd International Conference on Machine Learning, New York, NY, USA, pp. 2014–2023.
52. Gilmer, J., Schoenholz, S. S., Riley, P. F., Vinyals, O., & Dahl, G. E. (2017). Neural message passing for Quan-tum chemistry. In *International conference on machine learning*. Proceedings of the 34th International Conference on Machine Learning, Sydney, Australia, PMLR 70, 2017, pp. 1263–1272.
53. Li, Y., Yu, R., Shahabi, C., & Liu, Y. (2017). Diffusion convolutional recurrent neural network: Data-driven traffic forecasting. arXiv preprint arXiv:1707.01926.
54. Jain, A., Zamir, A. R., Savarese, S., & Saxena, A. (2016). Structural-rnn: Deep learning on spatio-temporal graphs. In *Proceedings of the ieee conference on computer vision and pattern recognition*. IEEE Explorer, Spain, pp. 5308–5317.
55. Yan, S., Xiong, Y., & Lin, D. (2018). Spatial temporal graph convolutional networks for skeleton-based action recognition. In *Proceedings of the 32nd AAAI conference on artificial intelligence*. New Orleans, Lousiana, USA.

2 Knowledge Graphs
Connecting Information over the Semantic Web

Garima Gujral and J. Shivarama

CONTENTS

DOI: 10.1201/9781003310792-2

2.1 INTRODUCTION

The information landscape over the semantic web has been growing at a fast pace. With large datasets being generated every instant, it is important that we get a unified view of heterogeneous and disparate pieces of data. Knowledge graphs provide an efficient approach for data governance and managing metadata. Heterogeneous data from unstructured data sources requires linkages to make sense and offer higher quality linked information. Knowledge graphs are built on the fundamental principles of the semantic web. The semantic web was designed by Tim Berners-Lee and aimed at organizing the World Wide Web. The web comprises the most heterogeneous, disparate, and decentralized data pool. The evolution of the semantic web has taken a different path and is based on linkages between datasets. In 2001, Tim Berners-Lee said that "The Semantic Web will bring structure to the meaningful content of Web pages, creating an environment where software agents roaming from page to page can readily carry out sophisticated tasks for users".[1] Knowledge graphs have the potential to disambiguate data silos and support a more linked and integrated approach. The existing data models have been transformed into semantic knowledge models consisting of taxonomies, ontologies, controlled vocabularies, thesauri. Knowledge graphs also serve as large databases but in an integrated manner that represents meaningful relationships. Knowledge graphs are a prerequisite for achieving semantic intelligence that can help discover facts from a particular domain.

The fourth industrial revolution caused a major shift towards the growing knowledge economy. Intellectual capital is considered the most essential resource.[2] Since the early 2000s, there has been a considerable change in how people and businesses perceive knowledge as the basic form of capital. With technology spreading its tentacles in every domain, knowledge goes hand in hand, and both technology and knowledge are the basic factors of capital. Knowledge lies at the core of the knowledge economy. In order to understand what knowledge is, we need to take a deeper look at how it originates.

Figure 2.1 depicts how data, when sorted, arranged, presented visually, and explained with a story, takes the form of knowledge. Knowledge can be both in explicit and implicit forms. The ideas, thoughts, inferences, and linkages that are interpreted by our mind take form of knowledge over the time. The implicit knowledge when put down and visualized takes form of explicit knowledge. Knowledge graphs helps the computer replicate the similar connections and linkages with disparate pieces of data and information spread out over the web as a human brain.[3] Upon querying, the knowledge maps it brings out the same set of results as curated in semantic networks and linked maps. Knowledge graphs can be curated to offer customized and domain specific knowledge and related search results.

2.1.1 Explicit and Tacit Knowledge

Knowledge can be classified as explicit and tacit, based on the form it is made available to us.

a) **Explicit Knowledge:** Knowledge available to us in the form of written material or published text such as books, research papers, articles, blogs, grey literature, etc. Explicit knowledge can be codified and digitized. It

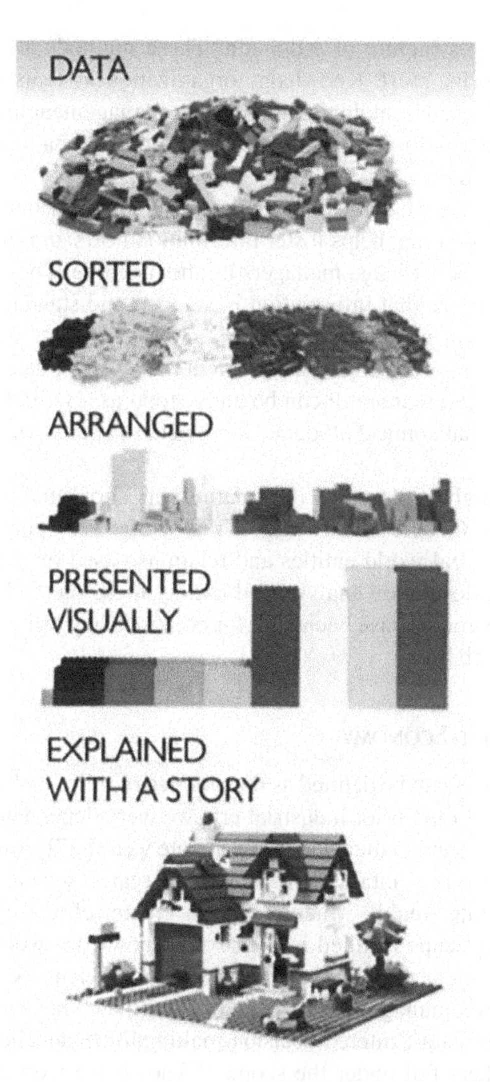

FIGURE 2.1 Data: sorted, arranged, presented visually, and explained.

is more objective in nature.[4] Transmitting and sharing explicit nature is easier than tacit knowledge. It can be archived and retrieved as a physical entity.

b) **Tacit Knowledge:** Knowledge embedded in the human mind through lived experiences. Tacit knowledge can be more subjective than implicit knowledge. Tacit knowledge is more personalized and harder to communicate. It exists in the form of ideas, thoughts, mindsets, and heuristics, and it carries values.

2.1.2 Knowledge Management with Knowledge Graphs

Organizing knowledge based on semantic knowledge model is a prerequisite for efficient knowledge exchange. Knowledge organization should be based on the underlying semantic structure of a domain. These methods are popularly used to classify and categorize stuff. Knowledge organization systems such as glossaries, thesauri, taxonomies, and ontologies support the management and retrieval of various forms of data and information. These systems make data machine readable and transferable for human use.

Knowledge graphs are capable for transferring cross-departmental and interdisciplinary communication that helps foster information flows and linkages. They replicate the expertise of knowledge managers by the machine. They provide controlled vocabularies and ensure that information is mapped and standardized.[5] Semantic matchmaking helps in networking data in a domain specific manner. Knowledge graphs have the potential to fully link and orient the layers of data and information in a very human-centered manner. It can be understood as a virtual data layer over the existing data seers that connect all data. Good visualizations offer discovery support and search analysis.

They serve as highly integrated information repositors that links heterogeneous data from domains. Google's Knowledge Graph is the most popular and prominent example of linking real world entities and relations based on graphs. They require keep domain and information analysis and mainly focus on the linkages. Variety of generic knowledge graphs have been built for both industry and academia that utilize the data over the web.[6]

2.1.3 Knowledge-Economy

Knowledge-Economy can be defined as the practice of utilizing knowledge as capital to generate wealth. In the prior industrial era, we were dependent on means of production via machines and manual labor to generate wealth.[7] However, this has been replaced by intellectual capital. Technology has created several new opportunities and doors to generate wealth, which results in greater efficiency and productivity. The knowledge landscape resulted in the rise of knowledge workers also known as knowledge management professionals or information scientists. They are the ones who archive, retrieve, manage, organize, and disseminate knowledge. They deal with in-depth nuances of data-centered decision-making.[8] In advanced economies over two-thirds of workers fall under the scope of knowledge workers. The knowledge economy functions hand in hand with human capital, intellectual capital, innovation, and creativity.

2.2 TRACING THE GROWTH OF SEMANTIC WEB

Time Berners-Lee, the founder of the World Wide Web in the year 2001, proposed the idea of the semantic web and mentioned it as the next big step: "The Semantic Web will bring structure to the meaning- full content of Web pages, creating an environment where software agents roaming from page to page can readily carry out sophisticated tasks for users".[9] Twenty years down the line the semantic web developed and grew and in a different direction. The World Wide Web Consortium (W3C) laid the blueprint for the semantic web as several recommendations:

- 1999: Resource Description Framework (RDF) Model
- 2004: Resource Description Framework (RDF) and RDF Vocabulary Description Language. RDF Schema
- 2004: OWL (Web Ontology Language)
- 2008: SPARQL Protocol and RDF Query Language
- 2009: Simple Knowledge Organization System (SKOS)
- 2012: OWL 2
- 2012: R2RML
- 2014: JSON-LD: JSON based serialization for Linked Data
- 2017: Shapes Constraint Language (SHACL)

W3C developed several semantic web standards which are used across several multi-dimensional domains and have led to technology adaptation. The Linked Open Data Cloud was manifested in 2006 as the first successful step toward the semantic web and since then the occurrence of linked data. The adoption of Schema.org has also been a major leap for the development of the semantic web. The vocabularies are widely used to power rich extensible experiences.

Labeled Property Graph (LPG) model was developed in 2010 by the group of Swedish engineers. It emerged out from an enterprise content management system. They are also referred as a stepping stone for knowledge graphs. Labeled Property Graphs have evolved across various database and as part of the query languages.[10]

Tracing the Evolution of Knowledge Graphs:

- 1736: Inception of the Graph Theory by Leonhard Euler
- 1976: The first paper on conceptual graphs published by John F. Sowa
- 1982: Knowledge graphs initiated in the Netherlands by C. Hoede and F.N. Stokman
- 1999: Resource Description Framework (RDF) Model was published
- 2001: The article "The Semantic Web" by Tim Berners-Lee, Jim Hendler, and Ora Lassila was published in the *Scientific American Magazine*
- 2006: DBpedia project was created
- 2012: The first knowledge graph was introduced by Google
- 2018: GQL Manifesto was published

2.3 KNOWLEDGE GRAPHS

Knowledge graphs can be defined as a graph structured knowledge base that stores interlinked entities, concepts, objects, related events, and situations. They focus on the connections between concepts and entities. Knowledge graphs emerge out of the semantic web lens and describe the relations between disparate entities between schemas, interrelations between entities organized in a graph that covers various topical domains. They can be understood as digital structures that represent knowledge as concepts and depict the relationships between them. They may or may not include an ontology that allows both humans and machines to understand and reason

FIGURE 2.2 Elements of knowledge graphs.[11]

the basis of contents. They are based on ontologies to make use of schema layers and allow inferences for information retrieval tacit knowledge rather than only for queries based on explicit knowledge. Figure 2.2 shows the elements of the knowledge graph.

It acquires information and integrates into an ontology and allows users to derive and walk-through new knowledge.[12] They are not just another form of knowledge visualizations; they are semantic networks which are a data. It represents models of knowledge using algorithms. Examples of knowledge graphs include Genomes, Google Knowledge Graphs, DBPedia, FactForge, and Wordnet.

2.3.1 Components of a Knowledge Graph

Knowledge graphs are comprised of data sources, data catalogs, and data consumers along with artificial intelligence models, graph databases, integration, and knowledge graph management. This architecture can be adapted and extended in various forms based on case-to-case use cases. They act as right database models and provide easy access to data for users and developers, bringing in data to the right format and supporting combination of environments and hybrids. Integration scenarios for systems architecture is also made available as a service in organization and integration of data should be standardized.[13]

Knowledge organization tools such as taxonomies and ontologies are the foundation for building intelligent search and discovery tools as they provide a schematic scheme and rich browse and search experience.[14] Taxonomies are the first step used for organizing knowledge and ontologies, and knowledge model form the next step. The transition from taxonomies to ontologies required more complex data models and query languages. It should begin with developing knowledge models that analyze content and metadata to meet the discovery requirements for end users. Knowledge organization systems are comprised of taxonomies, ontologies, classification schemas, and controlled vocabularies. The knowledge organization systems are explained in detail as follows.

2.3.1.1 Taxonomy

Taxonomies describe the entire vocabulary, controlled vocabularies, and terminologies and provides a roadmap to organize knowledge. The term taxonomy originated from ancient Greek; 'taxis' means 'arrangement' and 'Nomia' means 'methods'. In simple words, it can be understood as the method of naming and classifying. As per the established terms, controlled vocabularies have a common understanding as they are written in natural language. A taxonomy is a model of a small part of the world; it describes the things that exist in a domain and the relationship between those things. It can also be called a knowledge model or a machine-readable model of a knowledge domain. In a taxonomy, we define the entire structure of a domain.[15, 16] It is a more expressive knowledge model with clearly defined relationships between concepts with a hierarchical structure in most cases, such as a parent-child or broader-narrower term relationships. They are highly flexible in nature and grow with the needs of an enterprise. Several relationships can be derived from the taxonomies. They are a set of specific concepts, classes, and individuals in an ordered manner.

2.3.1.2 Ontology

Ontologies describe the meaningful relations and map the data linkages for a domain of knowledge such as projects, people, properties, description, relationships. Taxonomies act as a basis for developing relevant ontologies. An ontology is a formal naming and definition of the types, properties, and interrelationships of entities in a particular domain. In an ontology, we define the types, properties, and relationships of concepts in a thesaurus. Ontologies are a form of knowledge engineering and have a higher level of semantic relationships. Designing ontologies may include various kinds of vocabularies and their relationships. They define content based on relationships between class, properties, and relationship types consisting of named instances along with the concept of classes and individuals.

Ontologies give a new dimension to the flat structure of knowledge organization tools. SKOS can be extended and integrated into comprehensive knowledge graphs. Ontologies performs functions such as classification and defining more specific relations and attributes. They integrate data and map relational data models. It also gives out new relationships emerging out of existing elements.

Ontology management requires multidimensional approaches such as:

- The range of semantic expressivity
- Complexity of ontologies is wider than taxonomies
- They have a strong focus on axioms
- Acts as a basis for building expert systems and focused on the creation of a schema

Ontologies represent the backbone for formal semantics and there are several representations:

a) **Classes:** The description of an entity is also known as class hierarchy. While dealing with information it can be organized in several classes and each entity would fall under one class.
b) **Relationship types:** Relationship between independent entities tagged with information and nature of relationship describe as hierarchical, semantic or as a flat list.
c) **Categories:** Entities can be categorized on aspects of semantics is used for organizing information.
d) **Data sources:** Each domain of knowledge has a variety of content and data sources focused on specific areas. Data sources live disparately in silos.
e) **Graph database:** Graph databases contain comprehensive references in source systems and the relationships between those objects.

2.3.2 Knowledge Graph Life Cycle

The operational steps for the development of knowledge graphs the stages ranges from inventory, extraction, and curation along with transformation steps that link to enrichment. These life cycles are intertwined with each other as user loops. Figure 2.3 shows the life cycle of the Knowledge graph.

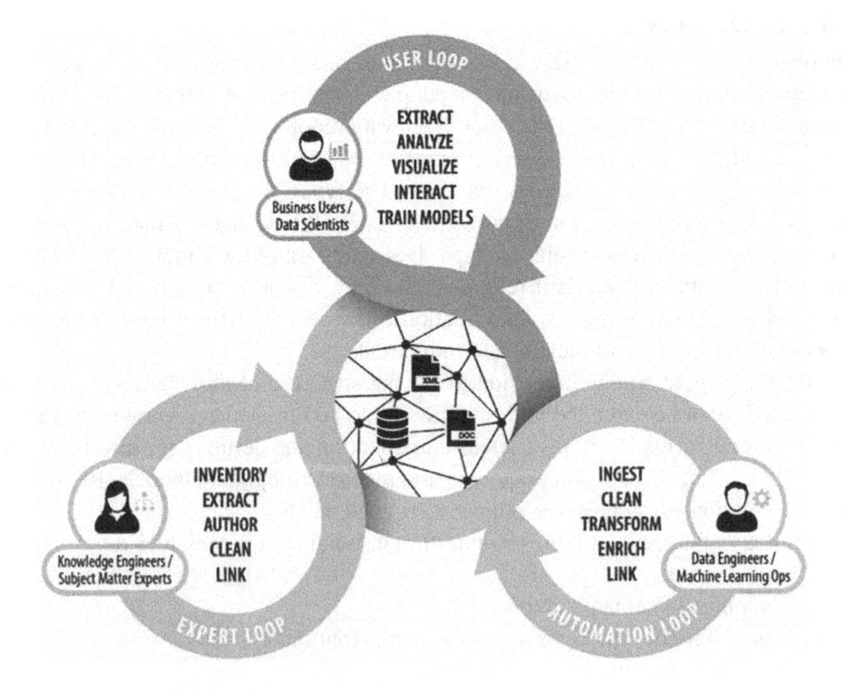

FIGURE 2.3 Knowledge graph life cycle.

(*Source:* The Knowledge Graph Cookbook)

These intertwined loops require several stakeholders and proceed in an iterative and agile manner. They are constantly linked together, and the primary aim is to balance the linkages, representing the domain knowledge and enriching it with user feedbacks.

These loops are explained in detail as follows:

a) **Expert loop:** The expert loops are designed by knowledge engineers and subject matter experts who work on taxonomies and ontologies the core tasks are inventory, extraction, cleaning, and linkages.
b) **Automation loops:** They are designed by data engineers and data operations for ingesting, cleaning, transforming, enriching, and linking data.
c) **User Loops:** The user loop was meant for beneficiaries to extract and analyze data, visualizing, and training models.

2.3.3 Steps to Build a Knowledge Graph

TABLE 2.1
Steps to Build a Knowledge Graph

SN	Step	Process
1	Identifying Needs and Requirements	Identifying the needs and requirements are the preliminary step towards designing a knowledge graph and understand what questions it is going to answer and the kind of target audience it intends to serve
2	Collecting and Analyzing data	Gathering the essential datasets and structures is the basis for a knowledge graph
3	Data Cleaning	Cleaning and filtering the datasets is important to remove all the clutter and meeting goals in an efficient manner
4	Creating Semantic Data Models	Upon thorough analysis of the data elements, the next steps is to create the semantic data models using standards such as RDF schema and OWL
5	Data Integration	Once the data has been converted into standardized knowledge models it needs to be integrated
6	Data Alignment	Aligning all the data elements is must with the ability to merge attributes and characteristics
7	Developing Data Architecture and Search Layer	Merging all the data elements to offer an enriched experience leading to more synchronized performance
8	Augmenting Graph based on Text Analysis	Enriching data and new entities from text to applying inferences to provide new information
9	Increasing Usability	The usability maximizes when the graph is able to answer all the questions for the end users and ensures the data is findable, accessible, interoperable, and reusable
10	Knowledge Graph Maintenance and Governance	Once the knowledge graph is ready the next step is making it easy to maintain and evolve as per the growing information needs of the users

2.3.4 Best Practices for Knowledge Graphs

The best practices for designing and maintaining knowledge graphs are as follows:

- Start small and grow gradually
- Prioritize and selecting use cases of implementing technical implementations
- Define the perfect data models for at every stage
- Use and apply personas to define the target audience and users of knowledge graphs
- Develop and test prototypes
- Work in an agile mode to manage data
- Get familiar with data and its forms
- Synchronize the data in a uniform manner
- Reuse and evaluate existing taxonomies and ontologies
- Create own independent ontologies as per the customized needs
- Apply existing knowledge organization tools
- Build on sold foundations
- Create a good URI scheme that are maintainable, consistent, and simple
- Ensure it is highly scalable and available
- Fully managed and cost effective for eliminating the need for hardware and software investments
- Query federation for knowledge graphs can be sourced from data across diverse sources and is stored natively

2.3.5 Knowledge Graph Services

Knowledge graphs perform several services such as

- **Ingestion services:** Allows connecting structured data from relational databases, unstructured data in systems, and content management platforms. Connectivity to APIs with external services to fetch linked data.
- **Consumption services:** Consumption services consists of data integration. Virtualization. Analytics and semantic AI. They should be made available via the application programming interface.
- **Orchestration services:** Orchestration services offer connectivity and modeling complex workflows and processes, logging delta execution, and processing for scheduling and limiting execution distribution.
- **Enrichment services:** Enrichment services are made available and sufficiently support the graphs by offering term extraction, tagging, named entity extraction, classification and fact extraction, rule-based extraction, entity linking, and content classification. They act as standards for taxonomies.

2.3.6 Popular Tools and Technologies for Building Knowledge Graphs

Knowledge graphs can be built using different technologies such as the World Wide Web Consortium (W3C) created on a set of technical specifications and query languages such as Resource Description Framework (RDF), SPARQL query language, and Web Ontology Language (OWL) that collectively make up the backbone of the semantic web. Semantic web was originated as the vision for how knowledge can be structured and managed early in 2000s when the beneficiaries laid the groundwork for the modern web.[17]

2.3.7 Graph Databases

Mapping and processing relations between entities and objects require graph databases. They are closer to the functioning of human brain and ways in which it generates meaning from the data. They are not the same as relational datasets but are designed as models consisting of nodes and representing entities and their relationships. They reduce the flexibility and agility of data modeling based on agile approach. RDF based graph databases are also called triplestores and labeled property graphs are used to develop knowledge graphs. Graph databases store data in the form of networked objects and the links between them. RDF triplestore are the popular choice for managing relational databases. They allow for adding metadata triples in a straightforward manner and have an enriched and contextual semantic relevance. RDF based graph databases are used to track data integration on a scale with interoperability frameworks.

2.3.8 Functions of Knowledge Graphs

Knowledge graphs perform several functions such as unifying data access, flexible data integration, and automating the process of data management. It creates higher level of abstractions and formats data in a unified manner. It is represented in a uniform, human friendly way that provides consistent view on diverse knowledge to decision makers with their need based on natural language processing. It is represented in a uniform and consistent view for interlinking scattered data across systems. Graph technology cerates richer semantic models and enhances augmented analytics. These descriptions have formal semantics and allows human and computers to process in an efficient and unambiguous manner. Diverse data is connected, and description of entities related to it. They combine characteristics of several platforms such as databases, graphs, and knowledge base because data bears formals semantics and is used to interpret and infer facts. These semantics enable humans and machine to infer new information without introducing factual into the datasets.

Knowledge graphs solve data management problems, such as multiplicity of data sources and types, represented at a higher level of abstraction and can deal with the diversity and solve issues with lack of centralized controls. It replicates the connective nature of how humans express and consume information by the computer and offers formal meaning that both machines and humans can interpret. The schema of data elements can be adapted, changed, and pruned while keeping it same. Knowledge graphs also serve as a knowledge base to hold the explosion of information in a well-organized manner. It helps find insights hidden in plain sight, aiding information discovery in regulatory documents, intelligent content, automated knowledge discovery, semantic search, and contextually aware content recommendation.

Analyzing unstructured and structured data together is processed by linking relevant pieces of data and building systems and people to projects connect related projects and centralize data. It improves process efficiency and analyzes the dependencies. Building virtual assistants, chatbots, reference, and question-answering systems to build context aware systems and answer queries from a vast knowledge base.

2.4 USE AND APPLICATION OF KNOWLEDGE GRAPHS

Knowledge graphs are used across several scenarios such as fabricating knowledge workflows, curating excellent customer experiences, connecting data for search and retrieval, and unifying disparate pieces of data. These areas of application are explained as follows:

2.4.1 Fabricating Knowledge Workflows in a Collaborative Manner

Classifying data and content in an integrated manner is essential to navigate information stored in different places across the web. Semantic based tagging is collaborative in nature and is widely used in digital asset management and content management systems. A term-based approach is already in use by various enterprises, which is being replaced by a graph-based approach. Tagging is the first step to organizing content, and it should take place simultaneously while content is being created. Tagging takes place in the background along with content production. These tagging workflows make good extensions to knowledge graphs. The search functions in systems can retrieve content based on tags. Semantic search functions such as facets, attributes, synonyms, and canonicals all can be linked together and form the elements of knowledge graphs. They are also used as search assistants and provide contexts for search results. Each digital asset leaves a semantic footprint. It allows to take places across systems, people, and projects all in a unified manner.

2.4.2 Unifying Structured and Unstructured Data

Unifying structured, unstructured, and semi-structured data eliminates data silos. Semantic knowledge graphs leverages the interrelationships between metadata. Mapping metadata systems meets requirements. Semantic data layers reduced metadata silos and make them explicitly available. This leads to finding and integrating data and collate semantic data models. These technologies can perform text mining and analytics across both structured and unstructured data.

2.4.3 Search and Retrieval

Large scale efficient search and retrieval networks are a must while dealing with enormous datasets and information spread across the web. Search queries ideally deliver results on the principles of precision and recall. Query languages such as SPARQL provide relational and navigational operators. Semantic searches offer a combination of high recall and precision, which provides results based on clustering based on heterogeneity. It tries to understand the intent of the search query and provides results that are close to the human understanding while linking all the relevant information. The results are delivered in an understandable manner for the purpose of fostering two-way dynamic interaction and faceted navigation. The front-end layer in the form of chat bots, AI bots, and virtual helpdesks have gained wide popularity. They have an enhanced level of natural language processing that is not just based on keywords but on concepts and their relations.

2.4.4 Web of Things (WoT)

Web of Things are also considered as a graph based on comprehensive knowledge models. These data models function based on deep analysis. Digital twins improve the quality of decision making.

2.4.5 Deep Text Analytic

Text analytic methods are not built on broad knowledge bases and do not incorporate domain knowledge models. Structured data can be matched and linked more easily and matched with other data sets. Deep text analytics uses knowledge graphs and semantic standards and can process the context of text being analyzed in a broader context. Several disciplines can be merged to efficiently develop semantic knowledge models. Unstructured communication and knowledge are generated because of human communication and deep text analytics can resolve ambiguity of the unstructured data. More precise recognition of human communication and transforming it is nature language processing (NLP). It is based on larger text units and prior knowledge.

2.4.6 Intelligent Robotic Process Automation

Robotic process automation (RPA) provides noninvasive integration with technology and is adopted by organization. They eliminate tedious tasks so that people can focus on tasks with higher value. RPA is widely used in tasks such as risk handling, helpdesk optimization, diligence processes, complaint, and claims handling processes.

2.4.7 Enhanced Customer Experience

Knowledge graphs provide and enhanced customer experience by creating a unified view of customer interactions and relationships. This enables customers with the right amount of information that allows them to make informed decisions. They provide a user oriented view of the content, which leads to connecting more economic opportunities and identifying the growing customer needs. Semantic footprints provide end users greater detail about the documents and products they are interested in.

2.4.8 Search Engine Optimization

Optimizing content to achieve the best possible search results involves a lot of strategies. Search engines when linked with knowledge graphs would give out more informed results with more context.

2.5 SEMANTIC KNOWLEDGE MODELING

Semantic knowledge modeling is very similar to how we humans tend to construct and understand knowledge of the world based on our experiences. Each person organizes information based on some fundamental principles, enlisted as follows:

a) Drawing a distinction between things
b) Uniquely naming the entities
c) Creating factual relationships between entities
d) Classifying things based on semantics
e) Creating general facts and relating the classes to each other
f) Creating formations and multilingual variations
g) Defining the contexts and framing entities based on a situation and adding a dimension\
h) Merging entities and keeping the trips
i) Inference and generating new relationships based on reasoning and facts

Triples form the core element of knowledge graphs they comprise of a subject, predicate and object with the data types such as Boolean, string, numerical, etc. The triples, along with labeled information, is better for readability. These human readable triples need to be made machine readable.

2.5.1 QUERYING KNOWLEDGE GRAPHS

Making structured data and unstructured data accessible in a unified manner is essential. Once the ontologies are set based on semantic standards and taxonomies with controlled vocabulary and language, they make structured data accessible in a connected manner. Retrieving data from different kinds of systems requires axes and insights. The SPARQL and RDF query language-based knowledge graph support accessing data and allows to query data and the context at the same time. They allow data integration and allows unknown data to be mixed. SPARQL queries are very expressive and allows validation of knowledge graphs. Graphs are represented in RDF serialization formats.

2.6 WAY FORWARD FOR KNOWLEDGE GRAPHS

With the onset of Covid-19, the information landscape and access pattern has completely transformed. Researching and searching solutions for a community requires organized data and the facts can be accessed, processed, and networked in a time bound manner. They are the basis for self-servicing and supporting two-way interaction channels for user benefits. While knowledge graphs are being widely used by human resource and knowledge organization systems and we have demonstrated examples of semantic AI and knowledge graphs used for the purpose of skimming through employee database in organizations. Linked open data is taken as a form to be recognized, invested, and based on added value of open data. They enhance data findability and have the potential to present data while limiting their openness that might made it hard or even impossible for reuse. For AI to work successfully, they need to make use of human knowledge and creativity and replicate it. This requires reworking AI architectures based on semantic design principles. The development of Graph Query Language has taken the direction to bridge the gaps with the existing semantic standards that offers benefits of approaches for upcoming new graph standards.

2.7 CONCLUSION

Knowledge graphs assist in meeting user needs and making critical decisions based on integrated knowledge model from various siloed source systems. They deliver scalability and enable responsive approach for data governance. They are transforming the entire information architecture and meeting the growing information needs of users. With the onset of Covid-19 followed by closures and lockdowns across the world require information professionals to meet user needs in a virtual manner eradicating the physical, time, and access restraints. Dealing with large amounts of diverse data present in the information landscape to meet user needs in a time bound manner can be managed with knowledge graphs in a unified manner. The data that an enterprise manages is too diverse, disparate, and present at large volumes. Insights and opportunities may get lost among the robust and tangled data pools that exist in siloed content.

Knowledge graphs provided a paradigm shift widely in industries and have had an enormous effect on systems and processes. Knowledge graphs use semantic metadata and provide a uniform and consistent view from diverse datasets, interlinking the knowledge that has been scattered across different sources and stakeholders. It captures the semantics of a particular domain and uses sets of definition, concepts, and properties along with the relations between them. The capture semantics of a particular domain are based on a set of definitions of concepts and their properties and relations along with logical constraints. They are built using a graph model and graph databases for the purpose of storing and navigating relationships with flexible schemas and provide highway performance for graph queries.

REFERENCES

1. Artiles, J., & Mayfield, J. (2012). Workshop on knowledge base population. In J. Artiles & J. Mayfield (Eds.), *Text analysis conference*. UK.
2. Carlson, A., Betteridge, J., Kisiel, B., Settles, B., Hruschka, E. R., & Mitchell, T. M. (2010). Toward an architecture for never-ending language learning. In *AAAI*, AAI, Berlin.
3. Chen, P., Lu, Y., Zheng, V. W., Chen, X., & Yang, B. (2018). Knowedu: A system to construct knowledge graph for education. *Ieee Access*, 6, 31553–31563.
4. Chen, Z., Wang, Y., Zhao, B., Cheng, J., Zhao, X., & Duan, Z. (2020). Knowledge graph completion: A review. *Ieee Access*, 8, 192435–192456.
5. Duan, Y., Shao, L., Hu, G., Zhou, Z., Zou, Q., & Lin, Z. (2017, June). Specifying architecture of knowledge graph with data graph, information graph, knowledge graph and wisdom graph. In *2017 IEEE 15th international conference on software engineering research, management and applications (SERA)*. IEEE, London, pp. 327–332.
6. Lehmann, J. et al. (2015). DBpedia: A large-scale multilingual knowledge base extracted from wikipedia. *Semantic Web Journal*, 6(2), 167–195.

7. Li, J., Zhou, M., Qi, G., Lao, N., Ruan, T., & Du, J. (Eds.). (2017, August 26–29). *Knowledge graph and semantic computing. Language, knowledge, and intelligence—second china conference (CCKS2017): Revised selected papers*, vol. 784. Chengdu, China: Springer CCIS.
8. Pan, J. Z., Vetere, G., Gómez-Pérez, J. M., & Wu, H. (Eds.). (2017b). *Exploiting linked data and knowledge graphs in large organisations*. Cham: Springer.
9. Lin, Y., Liu, Z., Sun, M., Liu, Y., & Zhu, X. (2015, February). Learning entity and relation embeddings for knowledge graph completion. In *Twenty-ninth AAAI conference on artificial intelligence*, AAAI Press, Palo Alto, California USA.
10. Pasca, M., Lin, D., Bigham, J., Lifchits, A., & Jain, A. (2006). Organizing and searching the world wide web of facts-step one: The one-million fact extraction challenge. In *Proceedings of the 21st National Conference on Artificial Intelligence (AAAI-06)*. Boston, Massachusetts.
11. Element of Knowledge Graphs, URL. www.ontotext.com/knowledgehub/fundamentals/what-is-a-knowledge-graph/. Online
12. Rotmensch, M., Halpern, Y., Tlimat, A., Horng, S., & Sontag, D. (2017). Learning a health knowledge graph from electronic medical records. *Scientific Reports*, 7(1), 1–11.
13. Singhal, A. (2012). Introducing the knowledge graph: Things, not strings. *Official Google Blog*.
14. Shi, B., & Weninger, T. (2018, April). Open-world knowledge graph completion. In *Proceedings of the AAAI conference on artificial intelligence*, vol. 32, no. 1. GHC, Nuremburg.
15. Xu, J., Kim, S., Song, M., Jeong, M., Kim, D., Kang, J., . . . Ding, Y. (2020). Building a PubMed knowledge graph. *Scientific Data*, 7(1), 1–15.
16. Lin, Y., Liu, Z., Sun, M., Liu, Y., & Zhu, X. (2015, February). Learning entity and relation embeddings for knowledge graph completion. In *Twenty-ninth AAAI conference on artificial intelligence*. Edeka, Vienna.
17. Zou, X. (2020, March). A survey on application of knowledge graph. In *Journal of physics: Conference series*, vol. 1487, no. 1. IOP Publishing, San Jose, p. 012016.

3 Latest Trends in Language Processing to Make Semantic Search More Semantic

Devi Kannan, Mamatha T. and Farhana Kausar

CONTENTS

DOI: 10.1201/9781003310792-3

3.1 INTRODUCTION

The field of question answering (QA) is a subset of Natural Language Understanding (NLU). It tries to develop systems that can extract important information from provided data and provide it in the form of a natural language answer when asked a question in natural language. For example, if you ask Alexa, "What is the schedule today?", she can pull raw information from the calendar about today's schedule and meetings. The response will be in the form of a statement in English. A computer's ability to understand natural language is defined as a programme which has the capability to translate phrases into an internal representation in order to provide valid replies to questions posed by a user.

Early QA systems for natural language comprehension [1] dealt how the natural language question and answer are processed and answer was communicated. Early conversations with a machine may have been text based or image based questions, questions in the form of list, inferential based, etc. In 1963,[2] a program called BASEBALL was developed to answer the questions from the stored database regarding the game of baseball. The information was stored in the list database, and based upon the syntactical analysis, the answer was searched and the result was displayed. Russell in his paper experimented image- and text-based information retrieval [3]. Despite the fact that the question is a text, this chapter presents an algorithm for obtaining an answer from an image. The previous question and answer systems are limited in scope and are not data-driven. They are mostly rule-based.[1–3] The Text Reception Conference (TREC) paved the way for finding answers to a question. This is regarded as the next generation of question-answering technology.[4] It shows how to give quick answers as well as the relevant documents for the query. Li and Roth (2002) introduced machine learning-based question classification.[5] This gives feature extraction from the question and gives accurate results using this learning approach.[6, 7] Before the advent of transformer architecture for transfer learning, unidirectional language models were commonly utilised, but they had various limitations, including reliance on unidirectional recurrent neural network (RNN) architecture and a short context vector size. To bridge these gaps and improve downstream task performance, bidirectional language models such as bidirectional encoder representation from transformer (BERT) are utilised. Natural language inference (NLI),[8] sentence-level paraphrasing,[9, 10] question answering (QA) systems, and token-level object recognition [7] are just a few of the tasks that bidirectional language models may help with. A probability distribution across word sequences is utilised to obtain information in the language model. N-grams, exponential neural networks (ENNs), and bidirectional models are examples of language models. Google Assistant uses these language models to evaluate data and predict phrases. BERT is the first deep bidirectional language model based on transformer architecture [11] that reads input from both sides, left to right and right to left, whereas earlier models were unidirectional and only read input from one side. BERT outperforms all other known models. This chapter is organized as follows: goals of question and answering, open domain architecture, open domain with retriever reader model, open domain with retriever generator model, and a conclusion. Each model is thoroughly explored, including its scope and limitations.

3.2 GOAL OF QUESTION AND ANSWERING

As a Google or Alexa end user, what do we anticipate from a search engine if we want to ask one question? We are looking for an exact solution to the question instead of receiving a long paragraph from which we must extract the response. For example, we asked the Google search engine "Who is the President of America". The answer is depicted in the Figure 3.1. However, sometimes we may occasionally receive the entire web page. So our goal is to create an intelligent system that extracts the exact answer rather than the entire page. In this chapter, we'll go over the many ways to provide an effective and relevant response to the proper inquiry.

The QA extract the reply for the questions from the unstructured database. The database may be from any websites: the organizational documents for the questions such as what, which when, how, etc. The inquiries could be of a domain-specific nature, such as free domain or closed domain. The free domain deals with anything, and in general it refers to database from any website or from any documents. The closed domain refers to a certain domain, such as questions relating to medicine or agriculture. In other words, the inquiries in this closed process are limited to a single area. The complexity of a closed domain is much lower than that of an open domain since it is dominated by a specific domain. In the remaining sections, we are going to discuss open domain only.

3.3 OPEN DOMAIN

An open domain question answering system in information retrieval that tries to provide a response to the user's query in the form of short text instead of a list of relevant papers/documents. In order to obtain this, the system uses several techniques such as language modelling, knowledge discovery, and formal specification approaches. When it comes to different forms of open-domain problems,[12] classifying it as model means that it must be able to extract the answer for the question either during the training time or the testing

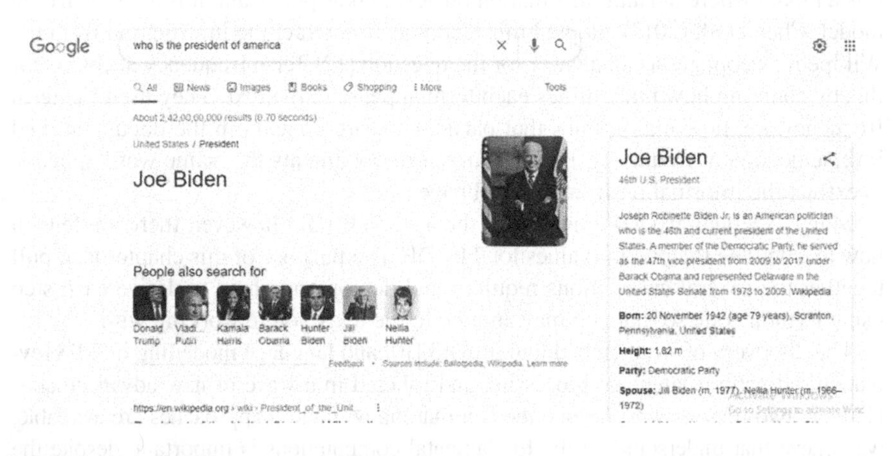

FIGURE 3.1 Google answer.

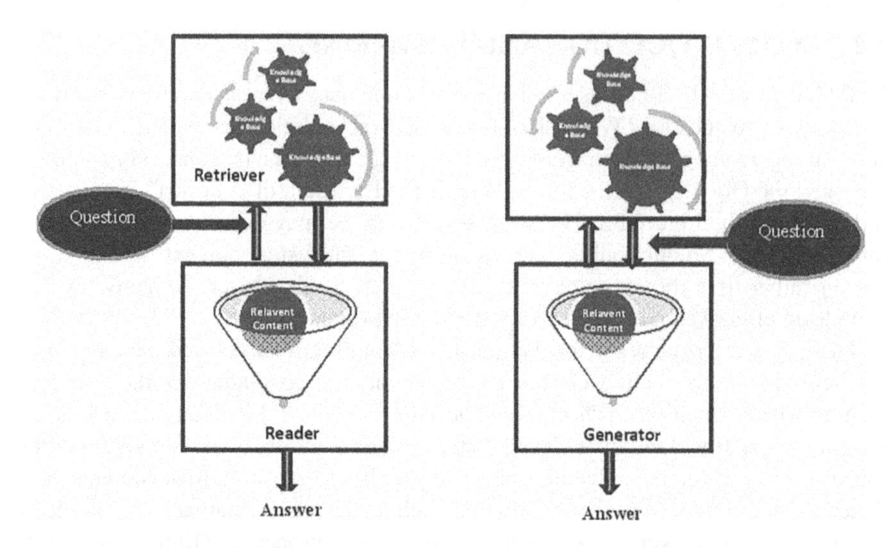

FIGURE 3.2 Reader and generator model.

time. Retriever Reader model is the name given to this scenario. It must also be taught in such a way that if the answer is not explicitly present in the training or testing data set, it can construct an answer from related documents. They call this a generator model. Figure 3.2 shows the reader and generator model.

3.4 OPEN DOMAIN MODELS FOR RETRIEVER READER

The open domain retriever reader model will extract the relevant document from the knowledge base as an answer to the question asked and display the result. There are several models are available for this. Most common one is the classic IR common information retrieval model Term Frequency Inverse Document Frequency (TF-IDF). It is a model where normal information retrieval takes place, and it is a non-learning model. Chen et al. (2017) adopted this technique to extract the information from the Wikipedia webpages as an answer for the question.[12] Term frequency accounts for this by counting how many times each term appears in the text. They used the term frequency, i.e. how many times that particular word appears in the document, and inverse document frequency, i.e. the number of documents that same word appears, to extract the information from that webpage.

Many excellent articles exist on the theory of TF-IDF; however, there are few on how to calculate the precise values for TF-IDF. The purpose of this chapter is to pull together the several calculations required and demonstrate how to derive each step using Python code so that you may apply it to the texts you're working with.

The discovery of word embedding in 2013 [10] and language modelling in 2018 revolutionised natural language processing and ushered in a wave of new advancements. Here are two reasons why, despite the fact that many modern approaches are available, we believe that understanding the fundamental computations is important, despite the fact that it was first proposed in the 1970s. One reason is that understanding the TF-IDF makes it easier to understand and assess the results of algorithms that use it. Google announced a text categorization method in 2018 that was built on 450,000 trials on a variety of text sets. When there were fewer than 1500 samples/words, Google discovered that TF-IDF was the best approach for expressing text. When you have a tiny sample size for a very common problem, the usefulness of TF-IDF is another reason to adopt this technique. The logic behind this technique is as follows.

Step 1: Create term frequency values from the ground up.
TF= Average of x in the document

Step 2: Create your own inverse document frequency values.
IDF= log(N/No. of documents that contain x).

Step 3: Using multiplication and normalisation, combine the above two numbers.
*TD-IDF = (TF*IDF)*

3.4.1 Retriever Reader Model

A source is a collection of information that are connected. In a typical natural language processing scenario, a corpus can span from a list of help desk files to a large range of scientific documents. We will express a few statements in this chapter to keep things simple.

For example:

Document A: There is a white cat on the wall.
Document B: The cat is white in colour.
Document C: People likes white cat.

Step 1: Token separation and calculate the term frequency
The first step is to separate the tokens in the given sentence and calculate the term frequency of each word. The TF is calculated as follows:
TF = (no. of repetitions of the word/total no. of words). For the tokens the TF is shown in the table.

TABLE 3.1
Calculate the Term Frequency

Word	There	Is	A	white	cat	on	the	wall	Total no. of words
Count	1	1	1	1	1	1	1	1	8

TABLE 3.2
Calculate the IDF

Word	There	Is	A	White	Cat	On	The	Wall	In	Colour	People	Likes
TD	0.125	0.125	0.125	0.125	0.125	0.125	0.125	0.125	0.166	0.166	0.25	0.25

Step 2: Calculate IDF

We seek a technique to reduce the relative impact of terms that appear far too frequently in all documents. Incorporate the inverse document frequency into your workflow. Naturally, if a word exists in all documents, it may not contribute significantly in distinguishing between them. The IDF is calculated as follows:

IDF = log(No. of documents/the number of documents that same word appears)

A term should be near the bottom of the 0–1 range if it appears in all of the texts. Using a logarithmic scale makes reasonable because log 1 = 0.

TABLE 3.3
Calculate the Term Frequency

Word	There	Is	A	White	Cat	On	The	Wall	In	colour	People	Likes
IDF	0.477	0.176	0.477	0	0	0.477	0.176	0.477	0.477	0.477	0.477	0.477

Step 3: Calculate TF-IDF

Term frequency (TF) and inverse document frequency (IDF) are combined in TF-IDF, which is determined by multiplying the two quantities together. The sklearn implementation then normalises the product between TF and IDF.

We can compute the TF-IDF in python using

TABLE 3.4
Computation of TF-IDF in Python with Example

```
vectorizer = TfidfVectorizer()
vectors = vectorizer.fit_transform([DocumentA, DocumentB,DocumentC])
cat colour in is likes on people
0 0.268062 0.000000 0.000000 0.345179 0.000000 0.453868 0.000000
1 0.300832 0.509353 0.509353 0.387376 0.000000 0.000000 0.000000
2 0.359594 0.000000 0.000000 0.000000 0.608845 0.000000 0.608845
 the there wall white
0 0.345179 0.453868 0.453868 0.268062
1 0.387376 0.000000 0.000000 0.300832
2 0.000000 0.000000 0.000000 0.359594
```

Tables 3.1 to 3.4 represent the TF-IDF computation for the Document 1, Document 2, Document 3 respectively. From this table, it is observed that cat, wall, and white are given more weightage in Document 1. However, colour, likes are not given more weightage in Document 1. From this, the retrieval reader will retrieve the answer from the document based upon the weightage.

3.4.2 Limitations and Hypotheses

The fundamental flaw in TF-IDF is that it ignores word order, which is crucial for comprehending the meaning of a phrase. In addition, record size might cause a lot of variation in TF-IDF readings. In the case earlier, stop words are not eliminated. Stop words can be eliminated in a variety of situations. I've merely used a small corpus for demonstration purposes. The corpus, vocabulary, and matrices illustrated in the picture are substantially greater in real-world use scenarios. This section includes a number of assumptions we followed.

3.5 OPEN DOMAIN MODELS FOR RETRIEVER GENERATOR MODEL

Instead of obtaining answers, the other types of QA generate them. The prior methodology extracted the answer based on the word's repetition or the relevance of the word count in that sentence. The meaning of the statement is another technique to get the answer from the question. The application of Transformer's bidirectional training to language modelling is an important technical advance for BERT. Previously, researchers looked at a text sequence from left to right or a combination of left-to-right and right-to-left training. In the realm of computer vision, transfer learning—pre-training a neural network model on a known task, such as Image Net, and then fine-tuning by using the trained neural network as the foundation of a new purpose-specific model—has been widely shown. A similar method has recently been proved to be effective in a range of natural language difficulties by researchers. Feature-based training is an alternative strategy that is also common in NLP tasks. A pre-trained neural network generates word embeddings,[10] which are then employed as features in NLP models in this method.

BERT mode is used to do this. BERT stands for Bidirectional Encoder Representation from Transformers, which is the well-known model utilised by Google to retrieve information from webpages. BERT is a deep learning model that uses Transformers to extract the response based on the weighted average of the answers to the question. The BERT model reads the material in both directions—whereas most models only read in one direction—and guesses the response using masked input and next sentence prediction approaches. The architecture of BERT will be discussed first, followed by its operation in the next section.

3.5.1 BERT and its Function

The input is converted to a numerical representation using this function (changing text into word embeddings.). BERT is a pre-trained model that may be fine-tuned for a variety of tasks. The BERT NLP Model was built using 2500 million words from

Wikipedia and 800 million words from literature as its foundation. Two modelling methodologies were taught to BERT:

Masked Language Model (Mlm)
Next Sentence Prediction (Nsp)

In practise, when employing BERT for natural language processing, these models are used to fine-tune text.

3.5.2 Model of Masked Language (MLM)

The masked language model (MLM) was briefly discussed in the previous section. MLM training is applied on a portion of the corpus that has been hidden or masked. BERT is given a series of words or sentences to form the sentence, and the contextual weights are maximised. BERT is given an incomplete statement and reacts in the most direct way possible. Consider the following sentence as an example:

In a year [Mask], summer, fall, and winter are the four [Mask] that alternate.

The [MASKS] keywords in the preceding text represent masks. It's a fill-in-the-blanks version.

Filling in the spaces will most likely reveal what completes the statement above. The statement's very first words after a year, we are able to decipher the sentence, and it becomes lot easier to comprehend. The term summer, winter is utilised to connect the phrase's numerous sections and serves as the final section's theme. This was simple to comprehend because we are familiar with English linguistic idioms and understand the contextual weights of these terms. To finish the statement, BERT must study extensively and grasp the linguistic patterns of the language. BERT might not know what summer or winter it is, but with so many words to process, the answers are almost surely spring and season.

To improve the output result, 15% of words in a sentence or text corpus are masked in practise. Two-thirds of a statement with 20 words can be concealed. The masked sentence can be created with the help of a masking function at the conclusion of the input. Remember that the goal of this approach is to improve the understanding of context between words in a phrase.

3.5.3 Next Sentence Prediction (NSP)

NSP aims to establish a long-term relationship between sentences, whereas MLM focuses on the interaction between words. According to Google's initial BERT NLP study, when NSP was not used, the model fared poorly across all measures. NSP entails giving BERT two sentences, Message 1 and Message 2 and IsNextSentence or NotNextSentence is BERT's response.

For example:

Message 1: I went to market and buy vegetables
Message 2: The vegetables are fresh
Message 3: It is so slow

Which of the messages do you think follows logically related from the other? Message 1 following Message 2 and Message 3 is almost certainly not. BERT is required to react to these questions. After MLM, teaching BERT with NSP improves the language's linguistic meanings and qualities to a bearable level.

3.5.4 Transformer Architecture in BERT Model

The BERT model is based on a Transformer architecture that includes encoder and decoder components. To decode the input based on the job, the encoder part consists of self-attention and feed-forward modules, while the decoder part comprises of self-attention, encoder-decoder attention, and feed-forward module.

3.5.4.1 Input to the Encoder

The Figure 3.3 shows the BERT input to the encoder. BERT receives a sequence of sentences as input for processing. As shown in Figure 3.3, this input sentence will be combined with two special inputs, CLS and SEP.

3.5.4.2 Token Embedding

In this token embedding phase the input statement is separated into tokens, and special symbols are added with the input statements. The CLS is a special symbol that appears at the start of a sentence, and the SEP is a sentence separator. The SEP is also used to indicate the end of a sentence.

Statement 1: **[CLS]** I Like coding **[SEP]**
Statement 2: **[CLS]** The students went to school. They play in the rain **[SEP]**

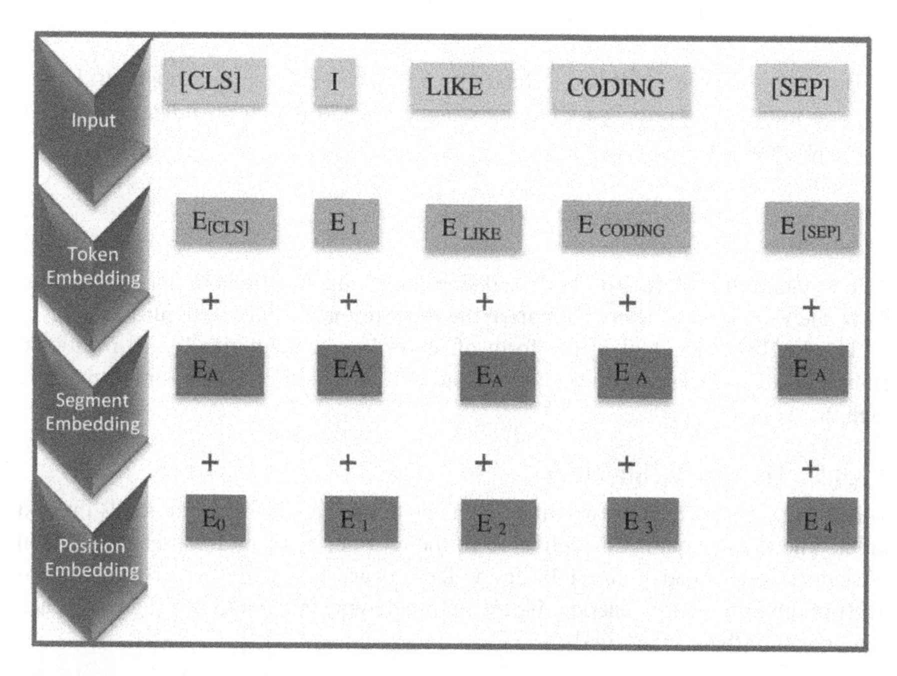

FIGURE 3.3 BERT input to the encoder.

```
text1 = "I Like coding"
Input_text1 = "[CLS] " + text1 + " [SEP]"
token_embedded_text = tokenizer.tokenize(Input_text1)
['[CLS]', 'i', 'like', 'coding', '.', '[SEP]']
```

3.5.4.3 Segment Embedding

This phase is used to add the segment as an indicator for each sentence. For example, here Statement 1 is added with segment E_A only, and in Statement 2 we have E_A and E_B as a segment embedding to distinguish the two sentences respectively. The segment embedding will be:

[CLS] "The" "students" "went" "to" "school" "." "They" "play" "in" "the" "rain" [SEP]
[CLS] E_A E_A E_A E_A E_A E_A [SEP] E_B E_B E_B E_B E_B [SEP]

3.5.4.4 Positional Embedding

The positional embedding shows the position of each token in the sentence. E_0 is for CLS. For Statement 1, the positional embedding will be:

Statement 1: **[CLS]** I Like coding **[SEP]**

E_0 E_1 E_2 E_3 E_4

But each token or word has different meaning based upon their position. For example, the words "park" in the sentence have different meaning at different positions.

She parks the car and play in the park

Ashish Vaswani et al. in 2017 [13] used sine and cosine functions to find the similarity in the words. So we can summarize the input to the encoder in Figure 3.4.

The English statement in the form of a word vector combined with positional embeddings produces a word vector of positional information, which is nothing more than the word's context.

3.5.4.5 Encoder Architecture

Transformer is made up of two different mechanisms: an encoder that reads the text input. The input sequence is delivered to the encoder block in parallel in the form of vectors. Multi-head attention layer and feed-forward layer are two crucial components of the Transformer encoder block. Before delving into detail, we'll go over each component of the encoder in detail.

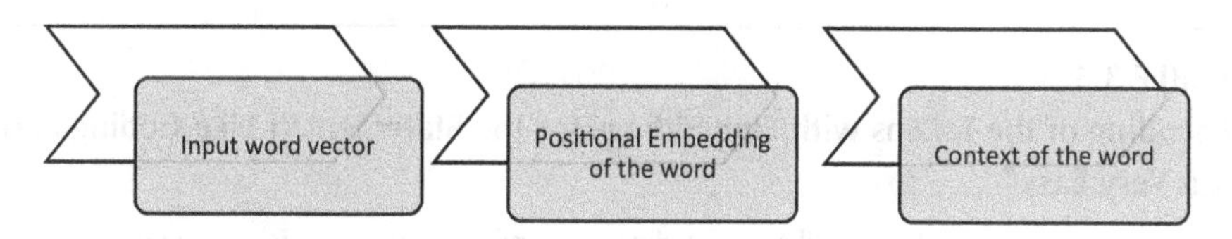

FIGURE 3.4 Input encoder.

3.5.4.5.1 Multi-Head Attention Block

The attention block is critical for determining the context of a word in a sentence. Before getting into detail, we'll go through the attention mechanism in detail. The component of the word on which we must concentrate is called attention. In other words, how relevant is the word in question to the other words in the sentence? This computation will yield the attention vector, which is nothing more than the contextual relationship between one word and another in the same sentence. For example the sentence "I like coding and it is very easy." What does it belong to in this statement? It doesn't matter if it's about I or coding. This refers to the word's context. The context of the word is determined by NLP. This portion is taken care by this attention block, and will discuss this in detail.

3.5.4.5.2 Self-Attention

Each word is handled as a token 't' when a machine reads this line, with a word embedding (V)WE for each token. However, there is no context for these word embeddings. The goal is to use some kind of weighing or something similar to generate a final word embedding (Y)WY that has more context than the initial embedding WE. Identical words appear closer together or have comparable embedding in an embedding space. For example, the terms 'bark' and 'breed' will be associated with the word 'dog,' rather than the word 'elephant.'

If the word 'student' appears at the start of the phrase and the word 'teacher' appears at the conclusion, they should provide each other greater context. To obtain more context, we can find the weight vectors W by multiplying (dot product) the word embeddings together in Table 3.5. Instead of using word embeddings like in the line, we multiply the embeddings of each word with one another.

So from the table, each word is to be multiplied with the other word to find the weightage. W11 to W18 are also normalised to a one-to-one total. The initial embedding of all the words in the sentence are then multiplied by these weights.

$$\text{Word embedding } Y_1 = \sum_{i=1}^{8} W_{1i}.V_i$$

The first word V1 appears in the context of all of the weights W11 through W19 in Table 3.6. As a result of applying these weights to each word, all other words are effectively reweighted in favour of the first. As a result, the word 'Bark' has become synonymous with the words 'dog' rather than the word that follows it. As a result, there is some context. This is done for all of the lexical items so that they all have some context.

TABLE 3.5
Encoding of the Tokens with Embeddings for the Statement "I Like Coding and it is Very Easy"

	I	like	coding	and	it	is	very	easy
Token	t1	t2	t3	t4	t5	t6	t7	t8
Word embedding (WE)	V1	V2	V3	V4	V5	V6	V7	V8
Weight	W1	W2	W3	W4	W5	W6	W7	W8
WE with context	WE	WE	WE	WE	WE	WE	WE	WE

TABLE 3.6
Computation of Word Embeddings and Normalization

Weight	V1.V1=W11	V1.V2	V1.V3	V1.V4	V1.V5	V1.V6	V1.V7	V1.V8
Normalized Weights	W11	W12	W13	W14	W15	W16	W17	W18

The fact that no weights are employed and that the order or proximity of the words has no influence on the conclusion is remarkable. Furthermore, the length of the sentence has no bearing on the procedure, implying that the number of words in a sentence is irrelevant. Self-attention is a technique for giving words in a phrase context.

Self-attention is problematic since it teaches nothing. If additional trainable parameters are provided to the network, it may be able to learn some patterns that provide much better context. This trainable parameter could be a matrix with trained values. The phrases "query," "key," and "values" were coined as a result.

Consider the previous statement once more where the initial word embeddings (V) are used three times. To get the weights, first make a dot product between the first word embedding and all other words in the phrase (including itself, 2nd), then multiply them three times to get the final embedding with context. The three Vs can be replaced by the phrases 'Query,' 'Keys,' and 'Values.'

Key	**V1**	**V2**	**V3**	**V4**	**V5**	**V6**	**V7**	**V8**
Query	V1. V1=W11	V1.V2	V1.V3	V1.V4	V1.V5	V1.V6	V1.V7	V1.V8
Normalized Weights	W11	W12	W13	W14	W15	W16	W17	W18

$$\textbf{Values } Y_1 = \sum_{i=1}^{8} W_{1i}.V_i$$

However, a matrix cannot be learned. That's all there is to it. As we know, multiplying a 1 x n shaped vector by a n x n shaped matrix yields another 1 x n shaped vector.

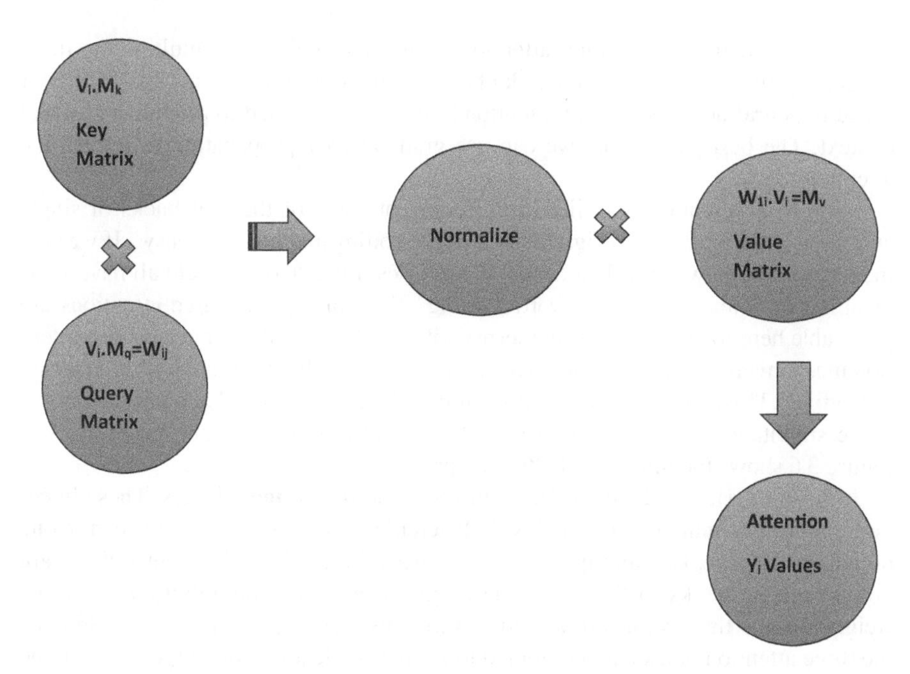

FIGURE 3.5 Self-attention process.

Keeping this in mind, multiply each key from V1 to V8 (each of shape 1 x n) by a matrix M_n (key matrix) of shape n x n. The values vectors are multiplied by a matrix M_v, and the query vector by a matrix M_q (query matrix; values matrix). The neural network can now learn all of the values in the Mk, Mq, and Mv matrices, offering far more context than just self-attention. Attention can be conceived of as a probabilistic version of information retrieval from databases. The whole self-attention process is shown in Figure 3.5.

Each of the key-value pairs k_i is compared to the query q using the attention approach. Each key value is given a weight based on its similarity. Finally, it generates a weighted combination of all the values in our database as an output. The sole difference between database retrieval and attention, in certain ways, is that database retrieval receives just one value as input, but attention receives a weighted combination of values. If a query is most similar to, example, keys 1 and 4, the attention mechanism will give these keys the highest weights, and the outcome will be a mix of value 1 and value 4. Using the query q and the keys k, the attention value is calculated, which is a weighted sum/linear combination of the values V. The weights are calculated by comparing the query to the keys.

3.5.4.5.3 Muti-Head Attention

The word embeddings have to go through a series of linear layers first. Because they lack a 'bias' term, these linear layers are effectively matrix multiplications.[13] The first layer is referred to as "keys," the second layer as "queries," and the third layer as "values." The weights are determined by multiplying the keys and queries in a matrix before normalising it. The final attention vector is calculated by multiplying these weights by the values. This block, dubbed the attention block, is now available for usage in neural networks. More attention blocks can be added to provide additional context. The best part is that we can use gradient back-propagation to refresh the attention block.

Multi-head attention is utilised to circumvent some of the drawbacks of single attention. Returning to the original line, "I like coding and it is very easy." If we take the term 'coding,' we may deduce that the phrases 'like,' 'easy,' should all have some meaning or connection to the word 'coding.' We can say that three attentions are preferable here to signify the three terms with the word 'coding' because one attention mechanism may not be able to accurately identify these three words as relevant to 'coding.' This eases the burden of locating all significant words on one's attention while simultaneously increasing the odds of finding more relevant terms quickly. Figure 3.6 shows the multi-head attention process.

To keys, queries, and values, let's add some additional linear layers. These linear layers all train simultaneously and with different weights. As a result, instead of one output, each value, key, and query now produces three. Three different weights are now available thanks to these three keys and queries. By multiplying these three weights by the three values in a matrix, three distinct outputs are created. Finally, the three attention blocks are combined to form a single attention output. We, on the other hand, chose three at random. In the actual world, they can be any number of linear layers, which are referred to as heads (h). That is, h linear layers can generate h attention outputs, which are then combined. That's why it's known as multi-head attention (multiple heads)

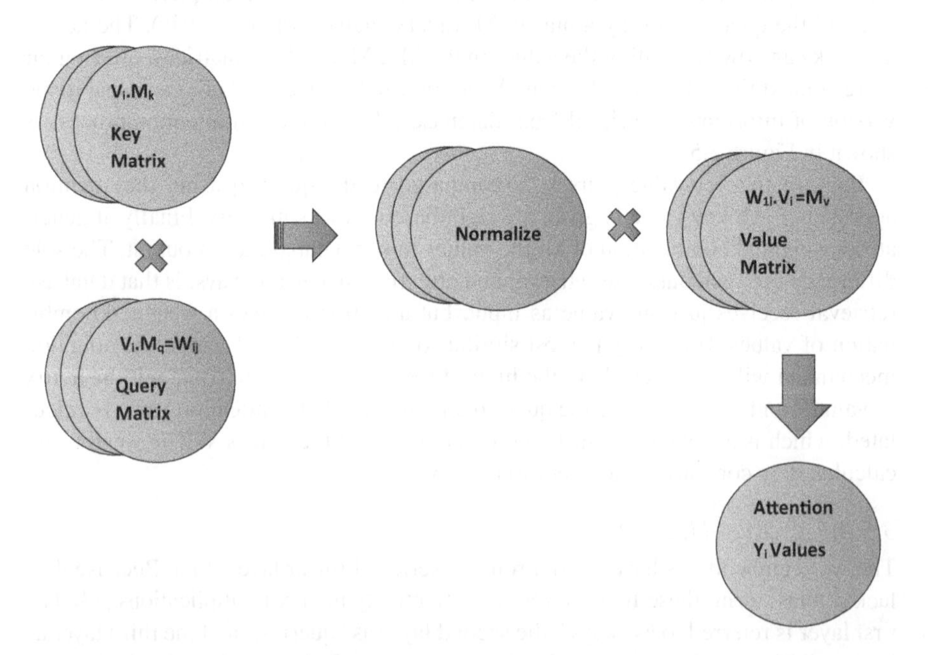

FIGURE 3.6 Multi-head attention process.

3.5.4.6 Decoder Architecture

The task of the decoder is to predict the given task. A self-attention mechanism, an attention mechanism over the encodings, and a feed-forward neural network are the three major components of each decoder. The decoder works similarly to the encoder, but with the addition of an additional attention mechanism that extracts useful information from the encoders' encodings.

3.6 QUESTION ANSWER IN BERT MODEL

The BERT model is similar to a blackbox. It will take the user's input/question and turn it into a token. In order to forecast the outcome, the BERT model hides 15% of the input and the computer cannot read characters, so it will turn the tokens into vectors first. To anticipate the answer, this input token was trained using an unsupervised learning pattern. We'll go over how the BERT model semantically retrieves the solution from the passage that follows. First, we'll require the paragraph where the solution will be found, and the query will be submitted to the model as an input.

Passage:

The president of the United States (POTUS)[A] is the head of state and head of government of The United States of America. The president directs the executive branch of the federal government and is the commander-in-chief of the United States Armed Forces.

The power of the presidency has grown substantially [ll] since the office's establishment in 1789.[6] While presidential power has ebbed and flowed over time, the presidency has played an increasingly strong role in American political life since the beginning of the 20th century, with a notable expansion during the presidency of Franklin D. Roosevelt. In contemporary times, the president is also looked upon as one of the world's most powerful political figures as the leader of the only remaining global superpower.[12–15] As the leader of the nation with the largest economy by nominal GDP, the president possesses significant domestic and international hard and soft power.

Question:

"Who is the head of state?"

3.6.1 Model Training

The preconfigured BERT packages are used to train the model. BertForQuestionAnswer and BertTokenizer are the predefined class available in Python.

```
from transformers import BertForQuestionAnswering
from transformers import BertTokenizer
```

```
model=BertForQuestionAnswering.from_pretrained('bert-large-uncased-
whole-word-masking-finetuned-squad')
tonkenizer_for_bert=BertTokenizer.from_pretrained('bert-large-uncased-
whole-word-masking-finetuned-squad')
```

BERT uses word embedding for training, as previously mentioned. As a result, the input passage and the inquiry must first be tokenized as shown below.

```
['[CLS]',
'who',
'is',
'the',
'head',
'of',
'the',
'state',
'[SEP]',
'the',
'president',
'of',
'the',
'united',
'states',
'(',
'pot',
'##us',
')',
'[',
'a',
']',
'is',
'the', . . . . . . . . [SEP]]
```

The tokenize checks first to see if the provided term is in the dictionary. It will store the term in the library if it is not available. Also in the above list some token consists of "##", it shows the sub-words of the token. This segmentation of tokens into subwords is done in order to give the word context and to improve performance.[13]

```
input_ids=tonkenizer_for_bert.encode(question,passage,max_length=max_
len,truncation=True)
```

The tokenizer encodes the token with the dictionary numbers, with 101 and 102 representing CLS and SEP, respectively.

```
[CLS] 101
who 2,040
is 2,003
the 1,996
head 2,132
of 1,997
the 1,996
state 2,110
[SEP] 102
the 1,996
president 2,343
of 1,997
the 1,996
united 2,142
states 2,163
(1,006
pot 8,962 . . . . . . . ]
```

For obtaining the answer to the question, the given question and passage are given training using the deep convolutional neural network layer.

```
question_scores=model(torch.tensor([input_ids]),token_type_ids=torch.tensor([segment_ids]))[0]
answer_scores=model(torch.tensor([input_ids]),token_type_ids=torch.tensor([segment_ids]))[1]
```

The question scores and answer scores columns represent the predicted values for the question and response, respectively. The values anticipated vary from negative to positive. The dissimilarity is shown by the negative indices. The higher the positive value, the more accurate the prediction.

```
answer_start_index=np.argmax(question _scores)
answer_end_index=np.argmax(answer _scores)
```

The anticipated answer for the question will have the highest value of the question and answer score. The expected response is

Answer:

"the president of the united states"

and the BERT model predicted the answer correctly.

3.6.2 Working with the BERT NLP Model: Things to be Taken Care

- The Word_Piece_tokenizer must be used when using BERT. You must use the same tokenizer that was used to train BERT to train your model.
- From the ground up with the BERT model instead of training fresh models, use pre-trained models when using BERT. This is quite expensive and not advised.
- The results from the runs will not converge when fine-tuning some jobs known as degeneration. This is primarily depending on the duties, and it is recommended that while fine-tuning it yourself, you be aggressive with early stopping.

3.6.3 Advantages of BERT Model

- BERT has been pre-trained on a significant amount of data and can recognise words based on their context.
- BERT is effective for task-specific models. The BERT model is easier to utilise for smaller, more defined jobs because it was trained on a large corpus. Metrics may be fine-tuned and implemented immediately.
- Because it is frequently updated, the model's accuracy is exceptional. This can be accomplished by careful adjustment.
- The BERT model is available and pre-trained for several languages. This is crucial for non-English projects.
- BERT is a free software project.

3.6.4 Disadvantage of BERT

BERT's size is responsible for most of its flaws. While using a large corpus to train data improves how the device anticipates and realises, there is another side to it. The disadvantage includes

- Because of the training framework and corpus, the model is quite huge.
- It takes a long time to train because it is so large and there are so many weights to update.
- It's not cheap; because of its vastness, it necessitates more computation, which comes at a cost.
- It must be fine-tuned for downstream activities, which may be picky because it is designed to be fed into other systems rather than being a separate programme.

3.7 CONCLUSION

By comprehending the content of the search query, semantic search aims to increase search accuracy. Semantic search, unlike typical search engines, can find synonyms as well as find items based on lexical matching. In this chapter, we spoke about different approaches for getting a user's answer semantically. The TF-IDF is one option.

The TF-IDF technique determines the importance of words based on their frequency in a document. It's a simple yet obvious system for weighing words, which makes it a good place to start for a variety of projects. Examples include creating search engines, document summaries, and other information retrieval and machine learning tasks. It's worth mentioning that TF-IDF can't help with semantic meaning transfer. It assesses the value of words based on their weighting, but it is unable to identify context or grasp relevance in this manner. Another problem is that, due to the curse of dimensionality, TF-IDF may suffer from memory inefficiency. Remember that the length of TF-IDF vectors equals the vocabulary size. While this may not be a concern in some categorisation scenarios, such as clustering, it may become unmanageable as the number of documents grows.

As a result, some of the alternatives, such as BERT, may be necessary, which is discussed in this chapter. Contextual knowledge of sentences has hampered natural language processing. Continuous improvement in this field will become even more precise in the future. All of these benefits are due to increased self-attention. This chapter simplifies BERT for easy understanding. To begin with, BERT is a precise, huge, veiled language model. Breaking down the technical jargon in the statement makes it easier to get a fast overview of what the model is about and what it strives to achieve. Its creation illustrates what happens behind the scenes, and question answering is an example of how it's used in practise. Chatbots, question-answering, summarization, and sentiment detection all appear to benefit from BERT and are other areas where BERT can be used to extract the information from the source. Unlike the TF-IDF model, which merely extracts the answer, one of the primary strengths of BERT is its contextual prediction of sentences. Another advantage is its prediction power. BERT also has several limitations, such as its ability to learn. Its ability to learn is determined by its size. When the corpus grows larger, the model slows down. The takeaway from this chapter is how semantically we can get the answer for the question in an efficient way. BERT is one such model which works semantically and efficiently depending upon the size of the database.

REFERENCES

1. Simmons, Robert F. (1965). Answering English questions by computer: A survey. *Communications of the ACM*, 8(1), 53–70.
2. Green Jr, Bert F., et al. (1961, May 9–11). Baseball: An automatic question-answerer. In *Papers presented at the western joint IRE-AIEE-ACM computer conference.*
3. Computer Interpretation of English Text and Picture Patterns Russell A Krisch, 1964 IEEE.
4. Voorhees, E., & Harman, D. (2000). *The Eighth text retrieval conference (TREC-8), special publication (NIST SP).* Gaithersburg, MD: National Institute of Standards and Technology. Retrieved May 28, 2022 https://doi.org/10.6028/NIST.SP.500-246 (Accessed May 28, 2022)
5. Li, Xin, & Roth, Dan. (2002). Learning question classifiers. In *COLING 2002: The 19th international conference on computational linguistics-Volume 1.* Association for Computational Linguistics, 1–7.
6. Voorhees, Ellen M., & Dawn, M. Tice. (2000). Building a question answering test collection. In *Proceedings of the 23rd annual international ACM SIGIR conference on research and development in information retrieval.* In SIGIR (SIGIR '00). ACM, New York, NY, USA, 200–207.

7. Mikolov, T. (2013). Distributed representations of words and phrases and their compositionality. *Advances in Neural Information Processing Systems*, 2, 3111–3119.
8. Bowman, S. R. (2015). A large annotated corpus for learning natural language inference. https://arxiv.org/abs/1508.05326
9. Williams, A., Nangia, N., & Bowman, S. R. (2017). A broad-coverage challenge corpus for sentence understanding through inference. https://arxiv.org/abs/1704.05426
10. Dolan, Bill, & Brockett, Chris. (2005). Automatically constructing a corpus of sentential paraphrases. In *Third international workshop on paraphrasing (IWP2005)* at International Joint Conference on Natural Language Processing (IJCNLP).
11. Devlin, J., Ming-Wei Chang, K. L., & Toutanova, K. (2018). BERT: Pretraining of deep bidirectional transformers for language understanding. arXiv preprint arXiv:1810.04805.
12. Chen, D., Fisch, A., Weston, J., & Bordes, A. (2017). Reading wikipedia to answer open domain questions. In *Proceedings of the 55th annual meeting of the association for computational linguistics (volume 1: Long Papers)*. Association for Computational Linguistics, Vancouver, Canada, pp. 1870–1879.
13. Vaswani, Ashish, et al. (2017). "Attention is all you need." *Advances in Neural Information Processing Systems* 30.

4 Semantic Information Retrieval Models

Mahyuddin K.M. Nasution, Elvina Herawati and Marischa Elveny

CONTENTS

4.1 INTRODUCTION

Information retrieval (IR),[1] from a philosophical point of view,[2] manifests as a science to reveal information-seeking behaviour,[3] while from a knowledge perspective, it acts as a technology to assess the relevance of information.[4] Suppose there is an information space and it is written as Ω, and there is information that is searched for or expressed as a search term t_x = '. . .'. Indirectly the information seeking behaviour reveals the relationship between t_x and Ω. On the implementation side, relevance assessment deals with the comparison between the information in Ω and the search term t_x, which is generally expressed as a percentage of the match between the two, or it is expressed as the relevance value $a\%$ in [0,1].[5–7] Furthermore, the value of relevance requires interpretation, namely a description that explains the meaning to determine the level of trust.[8] Therefore, this relation is recognized as

DOI: 10.1201/9781003310792-4

the semantic of IR, and of course, this requires a study of the patterns of the relationship between Ω and t_x.

In computational intelligence,[9] to deal with the exponential growth of information, it is necessary to transform human thinking into information systems that function to improve the performance of the relationship between Ω and t_x,[10] where the human mind recognizes detailed patterns with the constraint of slowing down time of processing as it increases data numbers. On the other hand, computers work relatively quickly but only against patterns that have been recognized, and they have decreased capacity for pattern learning.[11] Therefore, the similarity between the search terms t_x and ω in Ω becomes a challenge in estimating the value in [0,1]. It is referred to as a metric to measure semantic relatedness.[12] Then, with regard to data that is structured or unstructured, data science proposes data modelling methods for modelling the relationship between information space and search terms so that indirectly there are models of semantic information retrieval (SIR).[13–15]

4.2 RELATED WORK

Suppose the information space Ω consists of documents ω_i, i=1,2, . . . , I, where each document will consist of possible combinations of text, tables, and images. A document consists of one or more words, or there is a word set $W = \{w_k|k=1,2, \ldots, K\}$. The search term t_x consists of at least one or more words (also known as phrases), namely $t_x = \{w_1, w_2, \ldots, w_k\}$, and the size of the search term is the number of words in t_x or $|t_x| = k \geq 1$.[16] Naturally, words represent any object in the human mind that is expressed through the means of human articulation in the form of language.[17] Any word communicates any meaning to others in its community with the agreed meaning brought by the word. Then the word is encoded into characters such as a collection of letters and Arabic numbers, which are specially standardized and named the American Standard Code for Information Interchange (ASCII).[18] Although it is also possible that it involves the arrangement of other letters.[19] Therefore, the development of human knowledge as the object of human thought is transformed into documents, which are indirectly in the information space. When a word becomes common in its pronunciation, the word is a phenomenon before being registered in a language dictionary because the seeding of its meaning has not been completed in a semantic paradigm.[20]

4.2.1 Words from Data to Knowledge

Each word provides information in its meaning and then describes the object that the word refers to as knowledge.[21] Thus, indirectly, words become data, information, and knowledge and become part of human life without realizing it.[22] For that reason, the word refers to the naming of objects, activities, properties, each of which is expressed as a noun, verb, adverb, and which allows the presence of one word with another word having the same arrangement of letters. A word may be a derived word or a root word. This root word is called a token, and the token set is $\mathbf{w} = \{w_l|l = 1,2, \ldots, L\}$. Directly, it applies that $|W| \geq |\mathbf{w}|$ or $K \geq L$.

Definition 1. If there is the word w_k in ω_i, the evidence of w_k in ω_i is declared as the *occurrence* of w_k for $\|w_k\|_{\omega i} > 0$ as the number of the word events w_k in the documents ω_i, which is called the *hit count in singleton* of w_k.[23]

The parameter $\|w_k\|_{\omega i}$ in Definition 1 provides an estimated value of a word with the word probability formulation $p(w_k)$ in this document:

$$p(w_k) = \|w_k\|_{\omega i} / K \tag{4.1}$$

and by replacing $\|w_k\|_{\omega i}$ with $\|w_l\|_{\omega i}$, it results in an enhanced estimator: A probability

$$p(w_l) = \|w_l\|_{\omega i} / L \tag{4.2}$$

Equation 4.1 results in shows that if $\|w_k\|_{\omega i} \leq \|w_k\|_{\omega j} \leq \ldots \leq \|w_k\|_{\omega r}$ each is the hit count of the word w_k in the document ω, then the documents matched with w_k are ω_i, ω_j and ω_r in the appropriate order. Thus, there exists the document set $\Omega_{wk} \subset \Omega$, and $s_k = \|\Omega_{wk}\|$ is the cardinality of Ω_{wk}.[24] Similarly, Equation 4.2 gives the set of documents which may or may not be equal to the result of Equation 4.1 where $(s_k = \|\Omega_{wk}\| \leq (s_l = \Omega_{wl}\|)$.

Definition 2. If there is the word w_k, w_l in ω_i, the evidence of w_k and w_l in ω_i is declared as the *co-occurrence* of w_k and w_l for $\|w_k, w_l\|_{\omega i} > 0$ as the number of events of both words w_k and w_l in the documents ω_i which is called the *hit count in doubleton* of the words.[25]

Following Equation 4.1, the probability of two words $p(w_{km}, w_{kn})$ in one document is

$$p(w_{km}, w_{kn}) = \|w_{km}, w_{kn}\|_{\omega i} / K, \tag{4.3}$$

and following the change in Equation 4.1 into Equation 4.2 then to Equation 4.3 turns into the following equation:

$$p(w_{lm}, w_{ln}) = \|w_{lm}, w_{ln}\|_{\omega i} / L, \tag{4.4}$$

An equation is for increasing the estimate of co-occurrence.[26] Equations 4.3 and 4.4 can be expressed as symmetric conditional probabilities, namely $p(w_{km}, w_{kn}) = p(w_{km}|w_{kn}) = p(w_{kn}|w_{km})$. $\Omega_{wkm,wkn} \subseteq \Omega$ is the document cluster for w_{km} and w_{kn}, whereas $u_k = \|\Omega_{wkm,wkn}\|$ is the cardinality of the set, which is nothing but $\|w_{km}, w_{kn}\|$, where $\|w_{km}, w_{kn}\| \leq \|w_{lm}, w_{ln}\|$.

By ignoring the language structure or grammar in the document, the meaning of the implied sentences is accumulated into the explicit meaning by building relationships between words through symmetrical similarity measurements, for example, using the Jaccard coefficient.[27]

$$\begin{aligned} J_c &= p(w_{km}, w_{kn}) / (p(w_{km}) + p(w_{kn}) - p(w_{km}, w_{kn})) \\ &= (\|w_{km}, w_{kn}\|_{\omega i} / K) / ((\|w_{km}\|_{\omega i} / K) + (\|w_{kn}\|_{\omega i} / K) - (\|w_{km}, w_{kn}\|_{\omega i} / K)) \\ &= (\|w_{km}, w_{kn}\|_{\omega i}) / (\|w_{km}\|_{\omega i} + \|w_{kn}\|_{\omega i} - \|w_{km}, w_{kn}\|_{\omega i}) \text{in} [0,1] \end{aligned} \tag{4.5}$$

Since the alphabet came into writing to represent a thought, humans read and learn to gather information then interpret that information to develop science and technology or gain knowledge. Along with that, documents emerged to record and accompany human activities. The documents fill the shelves specially made for them. When the printing press was invented, it made printing and producing documents easier. As a result, academically, in addition to new documents emerging to tell the story of human life, it is easy to duplicate old documents. Libraries were built and developed with more robust shelves and to accommodate those documents. Those documents are often bound into books. For easy access, each book and shelf are assigned an identifier to be recorded on a card sheet (catalogue).[28] The library becomes an information space where appropriate information is arranged in such a way that users can easily reach that information. The catalogue system was the only one at that time that could recall the desired information. Likewise, in terms of language meaning, some words contain an inherent ambiguity. There are two fundamental reasons for semantic disambiguation of words to express general meaning, namely (a) different meanings can share the same word (*lexical ambiguity*),[29] and (b) one word can be assigned by many meanings (*referential ambiguity*).[30]

Access to information in the library is easier when an information technology appears and develops, where computer machines along with software provide a way to transform catalogues into a database, and where data becomes more structured and makes it easier to find information.[31] However, the formulation of the relationship pattern between any word w and document ω becomes more complex. The triplet model semantically states that any document consisting of a collection of words may describe the behaviour of the author, including the hidden intentions behind the writing of a document, and this reveals other information that needs to be crawled.[32] For example, the authors set is the set A = $\{a_n|a_1,a_2, \ldots ,a_N\}$ and suppose the default document set is called corpus $D = \{d_m|d_1,d_2, \ldots ,d_M\}$, then the form of the model aspect of $d \times a \times w$ or $\omega \times a \times w$ is as a triplet is as follows

$$P(\omega,a,w) = P(\omega)P(a,w\,|\,\omega) \tag{4.6}$$

for any document d in D or ω in Ω, any author a in D or a in Ω and a in A, and any word w in ω or w in Ω. The triplet states that there is a complete relationship between the three parameters ω, a, and w,[33] as shown in Figure 4.1.

For societal purposes, not long after the introduction of computers, the network of the Internet and the web completed the capacity and capabilities of digital technology so that the distribution of information could reach deep into the heart of different communities.[34] Anywhere that terminals are connected to the network they can obtain and send information to available storage places as a repository or online library. This convenience becomes the point of separation between the concepts of healthy and unhealthy reasoning for the behaviour of individuals, communities, or groups of people when disseminating information.[35] There are various efforts to improve the quality of human life and develop knowledge through the dissemination of information that is useful learning, but it can also lead to other destructive stories. Social engineering was carried out like sowing lies through perception of truth, but that perception was buzzed. Concepts and perceptions can be constructed involving

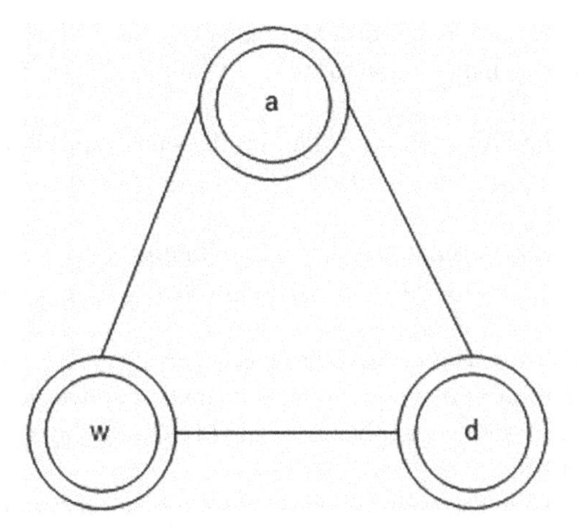

FIGURE 4.1 Triplet of a, w, and d.

the same media: text, data, databases, images, sound, text streams, sound streams, image streams (video), hyperlinks, etc. and are specifically divided into textual or non-textual or generally known as multimedia.[36] Like a flowing river, the flow of information is increasingly rushing against the walls of culture. Technology has become a double-edged sword for human well-being: one side upgrades it, the other side downgrades it. Therefore, the contents of the information space must not only be filtered, but also verified when dealing with heaps of data that are enormous in size, variety, etc. [37]

4.2.2 Search Engine

Both in terms of theory and in terms of implementation, the phenomenon of text as information accumulates into a paradigm called search engines.[38]

Definition 3. An information space Ω is declared as a search engine indexing the document ω_x if for each ω_x there is a search term t_x such that it forms the table $\Omega_x = \{(t_{xi},\omega_{xj})\} = \{(t_x,\omega_x)_{ij}\} \subseteq \Omega$.[39]

The search term is the content that is asked for answers from the information space, and the search engine provides a concise output of the possible information based on the estimated value of the search term from each document. The search term, in this case, is the content of the query which generally has a limit of k words or l tokens. Thus, the search term is presented as a vector of documents ("in/not in/of" web pages within the web) in the information space by any search engine with which the information space contains multi-dimensional semantics, namely:

1. *Meronymy* is a problem related to "is-a", part to whole relation or x is a part y. The semantic relation applies between part and whole.[40]
2. *Homonymy* deals with the issue of "has-a" or x has y, as part of itself, whole to part relation. The semantic relation that holds between a whole and its parts.[41]

3. *Hyponymy*, a subordination problem, namely x be subordinate of y or "has property". The semantic relation of being subordinate or belong to a lower rank or class.[42]
4. *Synonymy* is the problem that x signifies the same thing as y. The semantic relation that applies between two words or in context expresses the same meanings.[43]
5. *Polysemy*, a problem related to lexical ambiguity, separate words, phrases, or labels that can be used to express two or more different meanings.[44]

The search engine's treatment of the search term $t_x = \{w_1, w_2, \ldots, w_x\}$ is based on the assessment of truth of $\omega \rightarrow q$, a logical implication that applies to the query q where t_x in q and t_x in ω_x, i.e. $\omega \rightarrow t_x$ applies to ω in Ω based on the logical approach that $(\omega \rightarrow t_x) = 1$ (True).[45]

Theorem 1. *A document or web page is relevant to a query q if $\omega \rightarrow q$ is true for t_x in q.*

In other words, the probability t_x in power subsets is

$$P(t_x) = 1/2^{x-1} \tag{4.7}$$

thus giving a uniform mass probability function for Ω, P: $\Omega \rightarrow [0,1]$, where $\Sigma_\Omega P(\omega) = 1$. This reveals that the search term t_x forms the set of the singleton search term $\underline{S}$ based on Equation 4.7 and Definition 3, while a pair of the search terms t_x, t_y forms the set of doubleton search term $\underline{D}$ based on Definition 1 and Definition 2.

Lemma 1. If t_x in $\underline{S}$ is any search term that applies in the Ω information space as a search engine, then there is a document cluster Ω_x part of Ω as a representation of t_x or *singleton search engine event* of the information space.[46]

Lemma 2. If t_x, t_y in $\underline{D}$ any search term that applies in the Ω information space as a search engine, then there is Ω_{xy} subset of Ω as a representation of t_x, t_y or *doubleton search engine event* of the information space.[47]

$\|\Omega_x\|$ and $\|\Omega_{xy}\|$ are the cardinality of Ω_x and Ω_{xy}, respectively.

4.3 METHODOLOGY

The search term t_x in q which is fed into the search engine Ω serves to extract and generate information related to t_x from within Ω to the surface. The information derived from the document set Ω_x has characteristics corresponding to t_x.[48] When the search term t_x is came or derived from a set of documents as a dataset or corpus D, it directly establishes a relationship between D and Ω_x, i.e. $D \cap \Omega_x \subseteq D \cap \Omega$, which represents the probability of t_x,[49]

$$P(t_x) = \|D \cap \Omega_x\| / \|D \cap \Omega\|, \tag{4.8}$$

where $0 \leq P(t_x) \leq 1$, $\Omega_x \subseteq \Omega$ results in $\|D \cap \Omega_x\| \leq \|D \cap \Omega\|$ so that $P(t_x) \geq 1$. However, if the set of search terms $\{t_{xi}|i=1,2,\ldots,n\}$ is potentially derived from D, it is possible to generate a collection of document sets $\{\Omega_{xi}|i=1,2,\ldots,n\}$ such that $\cap_{i=1\ldots n}\Omega_{xi}\subseteq\Omega$, then for t_{xi}

$$P(t_{xi}) = \|D\cap\Omega_{xi}\| / \|D\cap\Omega\| \leq 1, \tag{4.9}$$

and

$$P(t_{xi} \mid i=1,\ldots,n) = \|D\cap U_{i=1\ldots n}\Omega_{xi}\| / \|D\cap\Omega\| \leq 1. \tag{4.10}$$

Equations 4.8 and 4.10 bear the relevance of the search term assignment results that match the expectations of search engine users by comparing the two lists of documents D and Ω_x, as a measure for IR.[50]

Definition 4. Suppose D is the document relevant to t_x, while Ω_x is the one retrieved by t_x, then the measurement of relevance is expressed as follows:

1. Precision is

$$Prec = \|D\cap\Omega_x\| / \|\Omega_x\|. \tag{4.11}$$

2. Recall is

$$Rec = \|D\cap\Omega_x\| / \|D\|. \tag{4.12}$$

Definition 4 reveals the importance of changing the structure of t_x or the content of the query q to overcome the problems encountered when dealing with large and stacked documents, varying both in form and source, in various types and velocities, including various patterns of relationships and meanings. Mathematically, the relationship between the search term t_x and the document ω_x is represented by all disambiguation problems which are formally summarized into two semantic tasks: (a) For some words w in W there is a relation ξ to assign a set Ω_w document containing w such that $\xi(1{:}N)\ W\rightarrow\Omega_W$, where there is w_i in Ω_w, $i=1,\ldots,n$ or ξ translates w to w_i in Ω_w i.e. $\xi(w) = w_i$ in Ω_x *(referential ambiguity)*.[51] (b) Suppose w_j in W, $j=1,\ldots,M$, there exists a relation ς to assign a set of documents Ω_W such that $\varsigma(M{:}1)\ W\rightarrow\Omega_W$, which has the word representative w in Ω_W, i.e. $\varsigma(w_i)$ = w in Ω_W *(lexical ambiguity)*.[52]

4.3.1 IR Formulation

The implications of the presence and meaning of words, search terms, search engines, and information spaces provide an opportunity to systematically formulate and study the terms of the context of words related semantically to deal with the dynamics of the information space.[33] Mathematically, the presence is expressed quantitatively, while the meaning is approached qualitatively. However, based on numerical investigation, a presence has the capacity to provide meaning, first time by weighting each search term [53, 54]

$$ft(t_x,\omega_i) = 1 \text{ if } t_x \text{ in } \omega_i, 0 \text{ otherwise} \tag{4.13}$$

where the *frequency of the word or term search* t_x in the document ω_i as the number of times t_x occurs in the document exactly expressed by,

$$tf(t_x,\omega_i) = ft(t_x,\omega_i)/\sum_{t'x in Éi} ft(t'_x,\omega_i) \tag{4.14}$$

However, the terms that appear frequently are not very informative. The *inverse document frequency* (IDF) is where there are a number of documents containing t_x, as a measure of the inverse of the non-informative of the word/search term t_x, i.e.

$$idf(t_x) = \log n / df(t_x) \tag{4.15}$$

n is the number of documents collected and $df(t_x)$ is the number of documents containing t_x, for example, in corpus or Ω_x. Therefore, numerically the words in the document are mapped as follows:

$$tf.idf = \left(ft(t_x,\omega_i)/\sum_{t'x in Éi} ft(t'_x,\omega_i)\right) \log n / df(t_x) \tag{4.16}$$

4.3.1.1 Keywords

It is possible that the potential recall of documents in more than one cluster is a result of the different semantic meanings of the search terms by index and weighting. Therefore, one model that changes the effect of indexing is by re-estimating numeric through the concept of co-occurrence in which keywords play a role in crawling any document related to the search term. For example, when an academic is looking for something by giving a query containing a name to a search engine, the information space may contain the same name with different people. Names that are not expected but are the same as names in t_x may be reduced by involving affiliates, for example. Keywords must be uniquely disclosed from the information space in latent.

The uniqueness of the keywords emphasizes the impression of the truth of the information disclosed. However, uniqueness is not easy to produce, there is a possibility that uniqueness comes from the language used. Another possibility is that a word or a group of words becomes an abbreviation that represents something uniquely. For example, USU is an abbreviation of "Universitas Sumatera Utara" (www.usu.ac.id/id) and is also an abbreviation of "Utah State University" (www.usu.edu/), but because the abbreviation comes from two different languages where the names of academics are generally different, USU maintains its uniqueness themselves to the names of academics.

Likewise, where the possibility of many semantic similarities between clusters of information space or classification, there are intersections of parts of the information space, and it is possible to generate more than one keyword, namely a set of search terms that become keywords. The effect of each of these search terms on the disclosure of information can be expressed as a comparison of the weights within the clusters or sections, $||t_{xk}||$ with the estimated value given by the information space, $||\Omega_{xk}||$, $k=1,2,\ldots,K$. The uniqueness of each word or search term from the keyword set is determined by the distance δ between $||t_{xk}||$ and $||\Omega_{xk}||$. The maximum distance between those distances determines the optimal keyword that can explore the depth of the information space, i.e. if $\delta_a \geq \delta_{k\text{-}a}$ then the best keyword is t_a, as shown in Figure 4.2, where

$$\delta = \left| \left(\|t_x\| / \sum_{k=1..K} \|t_{xk}\| \right) - \left(\|\Omega_x\| / \sum_{k=1..K} \|\Omega_{xk}\| \right) \right| \quad (4.17)$$

4.3.1.2 Information Space Semantic Topology

When the size of the data is so large along and has a wide range of information in a variety of domains, no matter how far the keyword goes, it will not actually result in all of the

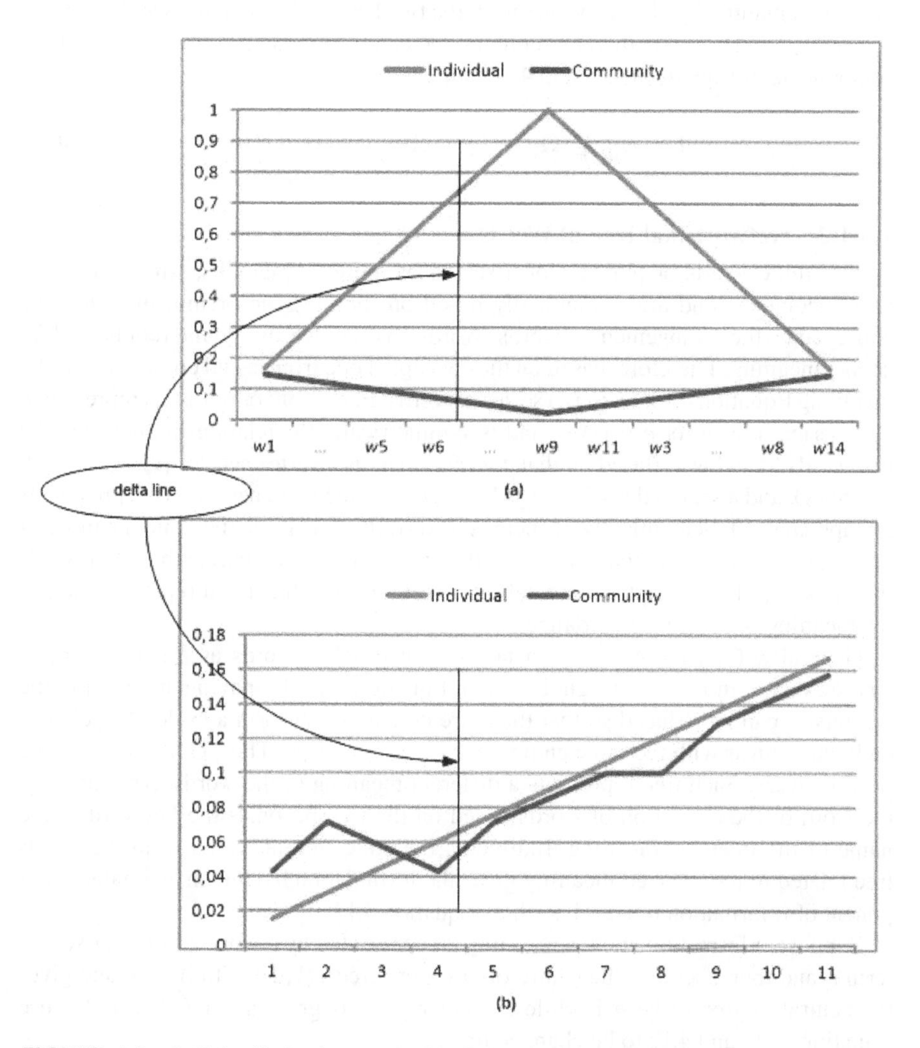

FIGURE 4.2 δ (delta line) of a word of individual or community: (a) set of words, and (b) word/years.

related information about something being disclosed. This reasoning leads to the need for a higher dimension of keywords. Involvement of some or all of the keywords is modelled with topological principles to get a broad range of these keywords.

Suppose the keyword set $\{t_{xi}|i=1, \ldots, k\}$ with each information space value of Ω_{xi}. The Ω_{xi} are parts of the information space Ω: $\Omega_{xi} \subseteq \Omega$. Suppose the universe of talk about the semantic space of information is expressed as τ, where $\emptyset$ (empty-set) and Ω are elements of τ, or $\emptyset$ and Ω in τ. Thus, the union of Ω_{xi} in τ with any sum, $k \geq 1$ for i, is a set X such that for all $X \subset \tau$, $\cup_{\Omega\ xi \text{ in } X} \Omega_{xi}$ in τ. The treatment of co-occurrence semantically gives the concept that a pair or more of Ω_{xi} in τ is in τ, i.e. for all n in N, for all Ω_{xi} in τ applies $\cap_{i=1 \ldots n} \Omega_{xi}$ in τ. Based on this, the reduction of the overlapping of keywords is possible to achieve the overall access to the information. Semantically, the co-occurrence, the two keywords t_x in pairs yields $er = \cup_{i=1 \ldots n-1} \cup_{j=i+1 \ldots n} (\Omega_{xi} \cap \Omega_{xj})$ thereby reducing the completeness equivalent to one of the components of Equations 4.11 and 4.12, namely

$$\Omega_x = \cup_{\Omega xi \text{ in} X} \Omega_{xi} - \cup_{i=1 \ldots n-1} \cup_{j=i+1 \ldots n} \left(\Omega_{xi} \cap \Omega_{xj}\right) \tag{4.18}$$

4.3.1.3 Network and Tree of Words

One or more words or phrases form sentences, build paragraphs, stretch message into documents, and are semantically based on the message or information to be conveyed by the arrangement of words. Words, in other words, are interconnected for deeper meaning. Therefore, based on the concept of occurrence and co-occurrence by involving Equation 4.10, there is a series of words. Each word or phrase is represented by a vertex, $v_i = w_i$, or $e_j = w_i w_{i+1}$ and is symmetrically the relationship between the two words w_i and w_{i+1}. Based on that concept, there is a vertex set $V = \{v_i|i=1, \ldots, n\}$, V is not $\emptyset$, and a set of edges $E = \{e_j|j=1, \ldots, m\}$. So the word network is expressed as a graph $G(V, E)$. Semantically, if there is a word that has more than one meaning or one name refers to more than one actor, then it is possible to separate the set of words that make up the description into subsets where each of them is a representation of the meaning or actor in accordance.

Generally, $G(V, E)$ represents a network of possible words as a perfect graph, where every e_j in E has a weight based on Equation 4.10. By paying attention to the weights, e_j can be reduced so that there are no edges that form a cycle. A cycle is a path that returns with the same ending: $e_a e_b, \ldots, e_j e_{j+1}, e_{j+1} e_a$. Thus, $G(V, E)$ forms one or more trees. Each tree represents a different meaning of the words. Semantically, the word or the collection of words based on the tree becomes the keyword of the name or the intended meaning. Indirectly, the collection of words from the words tree related to the referred meaning gives the formulation having an estimated component of information retrieval, such as Equation 4.17.

Suppose information about something is represented by a word, phrase, or search term t_x and then becomes the centre of any word tree $T(V, E)$ with $|V| = n$ and gives the central degree to be n-1 while the other point degrees are 1 (one), and cause Equations 4.11 and 4.12 to be changed to

$$Er,\ Prec = \left(\left\| D \cap \cup_{i=1 \ldots n-1} \Omega_{xi} \right\| / \left\| \cup_{i=1 \ldots n-1} \Omega_{xi} \right\|\right) - Er, \tag{4.19}$$

and

$$Prec = \left(\left\|D \cap \cup_{i=1\ldots n-1} \Omega_{xi}\right\| / \left\|D\right\|\right) - Er, \tag{4.19}$$

where *Er* is adjusted from the error model *er*.

4.3.1.4 Social Network

A social network is a model that relates a social actor to other social actors by using a graph as a representation of its structure where for a set of social actors $A = \{a_i|i=1, \ldots, n\}$, A is not Ø there is R as a collection of relationships between social actors based on the definition of any modality, namely a graph $G(A, R)$. Suppose the modality is expressed in the information space formed by a computer network with the principle that when a computer is connected to another computer and each computer represents the user as a social actor, then it is a social network, while the modality is the content of the computer network that accumulates in the space-server room or big data.

Extraction of social networks states that the modality of presenting social actors through occurrences using the names of social actors, while the relationship between two social actors is expressed through the co-occurrence of names of social actors. The trust level of a social network is modelled semantically and tested with a social network dataset based on the modality of the survey results, for example, the social network is $G(V, E)$. Thus, Equations 4.11 and 4.12, in order to express the truth of social networks, are based on a comparison between R and E, i.e.

$$Prec = |E \cap R| / |R| \tag{4.20}$$

and

$$Rec = |E \cap R| / |E| \tag{4.21}$$

On the other hand, information on a social actor can be disclosed from the information space by considering other social actors as keywords. The overall information of a social actor a_x in A is expressed by the degree of a_x in A,

$$\delta_{ax} = n - 1 \tag{4.22}$$

Based on the tree of $G(A, R)$, or a_x in A as the central vertex of a tree, as in Equation 4.18 in networks and word trees,

$$\Omega_{ax} = \cup_{\Omega ai \text{ in } A} \Omega_{ai} - \cup_{i=1\ldots n-1} \cup_{j=i+1\ldots n} \left(\Omega_{ai} \cap \Omega_{aj}\right), \tag{4.23}$$

gives the equation for precision and recall as follows

$$Prec = \left\|D \cap \left(\cup_{\Omega ai \text{ in } A} \Omega_{ai} - \cup_{i=1\ldots n-1} \cup_{j=i+1\ldots n} \left(\Omega_{ai} \cap \Omega_{aj}\right)\right)\right\| / \left\|\cup_{\Omega ai \text{ in } A} \Omega_{ai} - \cup_{i=1\ldots n-1} \cup_{j=i+1\ldots n} \left(\Omega_{ai} \cap \Omega_{aj}\right)\right\| \tag{4.24}$$

and

$$Rec = \left\| D \cap \left(\mathrm{U}_{\Omega ai \text{ in } A}\, \Omega_{ai} - \mathrm{U}_{i=1\ldots n-1}\, \mathrm{U}_{j=i+1\ldots n} \left(\Omega_{ai} \cap \Omega_{aj} \right) \right) \right\| / \| D \| \quad (4.25)$$

4.4 PROPOSED WORK

The roadmap for semantic information retrieval (SIR) starts from the concept of information retrieval in an article recorded in a Scopus reputed database (information space) entitled "Choice and coding in Information Retrieval systems" published in 1954,[55] which describes the trial of any storage (information space) containing a number of documents for retrieval of any information. It shows the importance of a framework called science to organize data so that information can be accessed easily, and the field of science is called information retrieval, which is a field of computers to search for a given set of data and extract records that meet a certain set of criteria.[56] [57]

4.4.1 Ontology and Taxonomy

Access to information naturally comes from human thinking about objects that humans can think of either through the help of the senses or through reflection, which humans then express in language, such as words, phrases, and sentences. Each word is a name for a tangible or intangible object: noun, work, characteristics, behaviour, and so on. It gives meaning to the word or gives birth to other words with similar or same meanings, synonyms. In this way, word by word, sentences are built to explain the possible meaning of any word, and it is seen in the (online)-dictionary as one of data sources, where every word has got an explanation. Ontologically, the existence of any object is due to the presence of instructions and data from the objects in the form of words, and then the presence of an explanation exists to interpret and function for life. For example, the object is "information retrieval" and can be described as "the activity of obtaining material that can usually be document on an unstructured nature i.e. usually text which satisfies an information need from within large collections which is stored on computers". The website1www.geeksforgeeks.org/what-is-information-retrieval/, access at June 2022. consists of a number of different words to describe the term "information retrieval". Likewise, in the online encyclopaedia Wikipedia, "information retrieval" also has the following description "The science of searching for information in a document, searching for document themselves, and also searching for the metadata that describes data, and for databases of texts, images or sounds",2https://en.wikipedia.org/wiki/Information_retrieval, access at June 2022. where each word has different weights as emphasis on meaning, for example, the word "searching" appears three (three) times. Thus, the existence of each object is evidenced by the number of words that name the object. For any search engine will present a hit count $\|\Omega_{object}\|$ when a query is submitted to a search engine to access an information space containing the term look for the named object.

Information ontology states that to show the existence of something as an object of thought, humans frame their meanings into definitions. For example, Table 4.1 provides several definitions of information retrieval. It shows the need for good

TABLE 4.1
Some Definition of Information Retrieval

	Description	Reference
1	A tool for ability to top the immense computational capacities of the computer.	[58]
2	An analysis (two-valued, many-valued and infinite- valued logic), and the truth-value is interpreted as the relevance value.	[59]
3	Conceptually fundamental in human communication as well as in man-computer communication.	[60]
4	How to decide, on the basis of clues, each of which is an imperfect indicator of document relevance, which documents to retrieve and the order in which to present them.	[61]
5	A list of technical terms with relations among them, enabling generic retrieval of documents having different but related keywords.	[62]
6	An interesting and challenging area of study and application, and it is an area currently in great flux.	[63]
7	The natural referent discipline for the study of model storage and retrieval and advance the argument that retrieving computer based models has an important connection to text retrieval based on documents' structure.	[64]
8	The recovery of documents that match a requester's query.	[65]
9	...	

interaction between information retrieval and Natural Language Processing.[65] The interaction produces a description of the role of words independently or together to extract information from the information space and through taxonomy the interaction is described, for example, in the form of a series of words, involving a number of computational completeness Equations 4.5, 4.10, 4.25, and 4.26. Using that concept, each row of the table represents a definition while the column of data is the same words used between definitions. The concepts of occurrence and co-occurrence produce words networks semantically and are reduced to words trees. Each word is a keyword, which semantically assigns the desired weighting of the extracting information from the stack of information spaces. By collecting definition after definition of information retrieval, collecting and upgrading semantic keywords can be a deterrent to accessing information related to documents about information retrieval.

4.4.2 Uncertainty

The concepts of uncertainty, such as probability, stochastic, and fuzzy, are innate traits of natural language. Words that describe the meaning or characteristics of quality—such as best, good, moderate, bad, worst—are words that refer to a level of estimation that cannot be clearly defined, except by assessing it in the framework [0,1]. The probability of a word, as in Equations 4.1 or 4.2, gives room to choose the one with the most potential based on the greatest probability. However, the specificity of the individual is formed from the distance between indeterminacy such as Equation 4.17. Therefore, based on the word order with the largest distance value δ_k, keywords are determined to access the most information. Likewise, word weights may result from the calculation of TF-IDF [16], which assigns appropriate weights to words or groups of words so that the uncertainty of the word as a keyword depends on the terms of the hit count based on the word and document in which the word belongs.

Directly or indirectly, by the fuzzy membership function

$$\mu_A(w):W \rightarrow [0,1], \tag{4.26}$$

fuzzy set A in W, for every w in W and $\mu_A(w)$ in [0,1]. By transforming the domain in W into [0,1] and involving the weighting of either Equations 4.1, 4.2, or 4.16, words sets can be classified into appropriate keyword sets in order to get the expected information. Directly, the fuzzy membership function gives an explanation of the interval [0,1] in possible parts that expressed by $0 \leq a \leq \ldots \leq b \leq 1$. The concept of a fuzzy membership function can also be applied to a strength relation where the weights are in [0,1] as a result of computing using Equation 4.5 to make it possible to decompose the word network into word trees that are suitable for certain information from an information space. However, each word tree becomes a marker for the information of a particular object that is estimated from the information space.

Dynamic changes occur in the information space with the reduction or addition of documents and can cause fluctuations in the probability of the word $p(w)$, which affects its role in the information space. Likewise, the same word w as a keyword has a variable role in the information space where relevance is sometimes ineffective. The fluctuating probability of the same word, called a stochastic indicator, semantically provides support for how to access the information space effectively. In other words, the comparison of the probabilities $p(w)$ and the resulting weights w will follow the concept of distance Equation 4.17, see Figure 4.2(a), for the same word different behaviour when following the time in which each word increases, see Figure 4.2(b).

4.4.3 Complementary

In particular, the information of each object has its own universe of discourse derived from its characteristics and behaviour as additional features of that object. This set of characteristics becomes an information subspace of the entire information space, where there are one or more other information subspaces. Information subspaces, therefore, have the possibility of being similar to one another, but each has the potential to be unique. This uniqueness is expressed as complementarity, i.e. the search term t_b has the following complement in the information space:

$$\Omega_b^{\,c} = \Omega - \Omega_b \tag{4.27}$$

which resulted in $\|\Omega_b^{\,c}\| = 1 - \|\Omega_b\|/\|\Omega\|$.

In an information space like the Internet, the web is a document with dynamics and the growth of the information subspace is always based on social actors. Thus, there are times when social actors have the potential to intentionally blow up certain information to cover up other information. This social actor is known as a buzzer, and it has a specific mission, such as sowing hoaxes to build certain perceptions. It results in the correct information not rising to the surface. Semantically, when the information space can recognize social actors as buzzers, incidental and co-occurrence searches with complementary filtering block information that is blown up so that hidden information can be pulled to the surface by keywords that are not t_b. For example, t_b in B, where B is the characteristic of all uniquely generated buzzers, then simply t_x is not in B as the search term used. However, decreasing the characteristics of buzzers requires separate tips involving the formulation and pattern of each buzzer.

4.4.4 Social Actors

The single most important entity, which has a strong influence on the growth of information in the information space where the exponential increase in documents continues, are social actors such as authors, editors, and so on. Humans are thinkers, researchers, learners, and social beings that generate documents that makes the information space continues to grow; after all, every document contains thoughts that are contained in the form of statements. Each statement requires evidence, such as statements about SIR like Table 4.2 or the statement "no research without publication", which has the potential to elicit responses from other parties like an article or paper that will be cited and commented on by different authors.

SIR by involving ontology questions about the creation, access, and representation of information directly related to humans is the issue of named entity recognition (NER), ambiguity in language, data duplication, and knowledge representation. The ontology has shown the existence of the information problem, not only through a problem statement but also by providing a solution with an appropriate method. The ontology is one of the methodology. Human limitations in spelling the problem has prompted a way to express it starting from the alphabet, rather than words, and then implanting human intelligence in search engines that work based on those words. While the search engine contains only the entire words w_k as a natural form created by humans to express world events, the search engine gives the probability of suitability using the concept of similarity using Equation 4.5. The search engine indexes

TABLE 4.2
Some Definitions of Semantic Information Retrieval

	Description	Reference
1	A hotspot of current research.	[66]
2	A method that more accurately narrows the searching scope and cuts down the redundancy.	[67]
3	An expansion technique that includes a mathematical model to compute similarity between concepts and an algorithm for query.	[68]
4	A model of information retrieval by using similarity calculation method.	[69]
5	A novel method based on a domain specific ontology.	[70]

all root words (token or vocabulary) w_l into itself and provide opportunities for conformance from Equation 4.5 for reaching to the source of information. With that concept, if the search term t_x consists of a root word, the artificial intelligence involved in all processes of accessing information gives a representation that $\|\Omega_x\|_{wk} \leq \|\Omega_x\|_{wl}$.

4.4.5 A Model of SIR

With that approach, however, it is semantically possible to increase the range of access to information by involving artificial intelligence that is ontologically derived from human thought. For example, following Equation 4.6, for any entity *a* in *A* and a number of keywords *x* in *X*, a mixture is generated based on the joint probability, i.e.

$$P(a,w \mid \omega) = \sum_{x \in X} P(x,w \mid x) P(x \mid \omega) \tag{4.28}$$

where the probability of the keyword *x* is $P(x|\omega)$, the probability of the word against the keyword is $p(w|x)$, and the probability of object for the keyword or key-phrase is $P(a|x)$. Suppose *x* in *X* is the centre point of the sequence between *a* in *A*, ω in Ω, and *w* in ω, i.e.

$$P(\omega,a,w) = \sum_{x \in X} P(x) P(\omega \mid x) P(w \mid x) P(a \mid x) \tag{4.29}$$

The posterior probability, the probability of an unknown quantity, estimates the keyword/key-phrase as the expectation of the observation (ω,*a*,*w*) which is computationally expressed as

$$P(x \mid \omega,a,w) \propto \left(P(x)P(a \mid x)P(\omega \mid x)P(w \mid x)\right) / \left(\sum_{x'} P(x')P(a \mid x')P(\omega \mid x')P(w \mid x')\right) \tag{4.30}$$

which makes it possible to maximize the expectations of the data environment, i.e.

$$P(a \mid x) \propto \left(\sum_{\omega,w} n(\omega,a,w) P(x \mid \omega,a,w)\right) / \left(\sum_{\omega,a',w} n(\omega,a',w) P(x \mid \omega,a',w)\right. \tag{4.31}$$

$$P(w \mid x) \propto \left(\sum_{\omega,w} n(\omega,a,w) P(x \mid \omega,a,w)\right) / \left(\sum_{\omega,a,w'} n(\omega,a,w') P(x \mid \omega,a,w')\right. \tag{4.32}$$

$$P(x \mid \omega) \propto \left(\sum_{\omega,w} n(\omega,a,w) P(x \mid \omega,a,w)\right) / \left(\sum_{\omega',a,w} n(\omega',a,w) P(x \mid \omega',a,w)\right. \tag{4.33}$$

where $n(\omega, a, w)$ indicates the number of occurrences of the word *w* in the document ω with the name *a*, sign ' in *a'* indicates complementary of *a*.

Equations 4.31, 4.32, 4.33, and 4.34 aim to reveal topics related to social actors with descriptions such as Table 4.3. Each topic of discussion has a different description based on the available w_i, i=1, . . . , n, but this uniqueness is also accompanied by the words w_j, j=1, . . . , m, which are descriptions between all topics differently. Thus,

TABLE 4.3
Generation of Author-Topic Relationships

Topic 1 "Keyword-1"	Topic 2 "Keyword 2"	. . .	Topic *n* "Keyword-*n*"
w_{11} . . . in [0,1]	w_{21} . . . in [0,1]		w_{n1} . . . in [0,1]
w_{12} . . . in [0,1]	w_{22} . . . in [0,1]		w_{n2} . . . in [0,1]
w_{13} . . . in [0,1]	w_{23} . . . in [0,1]		w_{n3} . . . in [0,1]
. . . . . . in [0,1]	. . . . . . in [0,1]		. . . . . . in [0,1]
w_{1m} . . . in [0,1]	w_{2m} . . . in [0,1]		w_{nm} . . . in [0,1]
a_1 . . . in [0,1]	a_1 . . . in [0,1]		a_1 . . . in [0,1]
a_2 . . . in [0,1]	a_2 . . . in [0,1]		a_2 . . . in [0,1]
. . . . . . in [0,1]	. . . . . . in [0,1]		. . . . . . in [0,1]

TABLE 4.4
Recall and Precision Based on Author-Topic Relationships

Topic "Keyword"	Precision	Recall	*F*-measure
w_{11} . . . in [0,1]	. . . in [0,1]	. . . in 0,1]	. . . in [0,1]
w_{12} . . . in [0,1]	. . . in [0,1]	. . . in 0,1]	. . . in [0,1]
w_{13} . . . in [0,1]	. . . in [0,1]	. . . in 0,1]	. . . in [0,1]
. . . . . . in [0,1]	. . . in [0,1]	. . . in 0,1]	. . . in [0,1]
w_{1m} . . . in [0,1]	. . . in [0,1]	. . . in 0,1]	. . . in [0,1]
a_1 . . . in [0,1]	. . . in [0,1]	. . . in 0,1]	. . . in [0,1]
a_2 . . . in [0,1]	. . . in [0,1]	. . . in [0,1]	. . . in [0,1]
. . . . . . in [0,1]	. . . in [0,1]	. . . in [0,1]	. . . in [0,1]

if the actors are also different, it is indirectly possible to involve the concept of similarities from Equation 4.5 or others similarity metric resulting in a social network between social actors based on different topic descriptions.

This social network is based on the strength of the relationship between topics. Likewise, the author has a role in each topic, so that each description has the potential to be a keyword other than the stated topic. Naming appropriate information is done semantically with the meaning of the topic and its description of the knowledge of the author (Table 4.4). Similar conditions are true for any social network.

Computationally, the IR measurement can be shown by the estimation of recall, precision, and *F*-measure of Table 4.4. The computation is based on Equations 4.19 and 4.25 with Equation 4.11 to guide for precision, and Equations 4.20 and 4.26 with Equation 4.12 to guide for recall, and

$$F - measure = (2 * Prec * Rec) / (Prec + Rec). \tag{4.34}$$

A metric used for imbalanced classification problems.

4.4.6 An Algorithm

Suppose there is a set of search terms, $T = \{t_{xi}|i=1, \ldots, n\}$, where $|T|$ is the size of T. Each t_{xi} is assigned to reveal the appropriate information via the query q that the search engine executes from the information space. The search engine indexes the documents ω into the contents of the information space Ω, or ω in Ω. The search term t_{xi} results in (a) clustering of documents Ω_{xi} which will consist of a summary of documents ω_{xl} together with identity id_l, or call it as a snippet; (b) hit count $||\Omega_{xi}||$ which represents the number of documents containing t_{xi}. So $||\Omega_{xi}||$ is the number of documents as the cluster of t_{xi}.

For example, a standard document set D_s is generated such that each document d_s in D_s has id_j, $j=1, \ldots, |D_s|$ where $|D_s|$ is the number of documents in D_s. By comparing the two documents, IR measurements can be organized (Algorithm I).

Algorithm 1:

1. $S \leftarrow D_s = \{(id_s, d_s)\}$
2. $a = 0$
3. $b = 0$
4. $T = \{t_{xj}\}$
5. for $j = 1$ to $|T|$
6. $q = t_{xj}$
7. $D_l \leftarrow \Omega_{xj} = \{id_j, \omega_{xj}\}$
8. $h \leftarrow ||\Omega_{xj}||$
9. for $I = 1$ to h
10. if id_i in S then
11. $a = a+1$
12. $b = b+1$
13. remove id_s
14. else
15. $b = b+1$
16. end if
17. end for
18. $S \leftarrow D_s = \{(id_s, d_s)\}$
19. end for

By using the computation results that are accommodated in the variables a and b, the value is $Prec = a/b$ or $Prec = a/||\Omega_x||$ and $Rec = a/||D_s||$. Likewise, the effectiveness of the b value can be increased by placing an instruction after step 18 $L = D_l$, and then after step 7 involve several instructions that compare the documents in D_l and L and discard the documents with the same id_j of D_l.

4.5 CONCLUSION

SIR is becoming something important, but it conceptually requires guidance in building theory and implementation. The guide is a definition based on the definitions of Table 4.1 and Table 4.2, as follows: As a current research hotspot, Semantic Information Retrieval (SIR) is a scientific field that involves methods based on the ontology domain, which is an extension of information retrieval (IR) models and techniques, through the involvement of computational intelligence and based on mathematics such as similarities where occurrence and co-occurrence play a role in determining the relationship between parts of the information space.

REFERENCES

1. Perry, J. W., Berry, M. M., Luehrs, F. U., J. R., & Kent, A. (1954). Automation of information retrieval. In *Proceedings of the Eastern joint computer conference: Design and application of small digital computers*. AIEE-IRE, pp. 68–72. https://doi.org/10.1145/1455270.1455287
2. Sievert, M. E. (1987). Information retrieval in philosophy. *Proceedings of the ASIS Annual Meeting*, 24, 254.
3. Milojević, S., Sugimoto, C. R., Yan, E., & Ding, Y. (2011). The cognitive structure of library and information science: Analysis of article title words. *Journal of the American Society for Information Science and Technology*, 62(10), 1933–1953. https://doi.org/10.1002/asi.21602
4. Smith, S., Jackson, T., & Adelmann, H. (2007). Concept clouds—Improving information retrieval. In *Proceedings of the European conference on knowledge management*. ECKM, Europe, p. 931.
5. Boisot, M. H., MacMillan, I. C., & Han, K. S. (2008). *Explorations in information space: Knowledge, actors, and firms*. Oxford: Oxford University Press, pp. 1–240. https://doi.org/10.1093/acprof:oso/9780199250875.001.0001
6. Boisot, M. H. (2013). Information space: A framework for learning in organizations, institutions and culture. In *Information space: A framework for learning in organizations, institutions and culture*, vol. 2, pp. 1–550. https://doi.org/10.4324/9780203385456
7. Nasution, M. K. M., Syah, R., & Elfida, M. (2018). Information retrieval based on the extracted social network. *Advances in Intelligent Systems and Computing*, 662, 220–226 https://doi.org/10.1007/978-3-319-67621-0_20
8. Bordogna, G., Carrara, P., & Pasi, G. (1992). Extending Boolean information retrieval: A fuzzy model based on linguistic variables. *International conference on fuzzy systems*. FUZZ-IEEE, San Diego, CA, USA, pp. 769–776.
9. Veningston, K., & Shanmugalakshmi, R. (2014). Computational intelligence for information retrieval using genetic algorithm. *Information (Japan)*, 17(8), 3825–3832.
10. Yang, H., Sloan, M., & Wang, J. (2015). Dynamic information retrieval modeling. In *WSDM 2015 — Proceedings of the 8th ACM international conference on web search and data mining*, pp. 409–410. https://doi.org/10.1145/2684822.2697038
11. Zhalehpour, S., Arabnejad, E., Wellmon, C., Piper, A., & Cheriet, M. (2019). Visual information retrieval from historical document images. *Journal of Cultural Heritage*, 40, 99–112. https://doi.org/10.1016/j.culher.2019.05.018
12. Jiang, Y. (2020). Semantically-enhanced information retrieval using multiple knowledge sources. *Cluster Computing*, 23(4), 2925–2944. https://doi.org/10.1007/s10586-020-03057-7
13. Nasution, M. K. M., Sitompul, O. S., & Nababan, E. B. (2020). Data science. *Journal of Physics: Conference Series*, 1566(1). https://doi.org/10.1088/1742-6596/1566/1/012034
14. Nasution, M. K. M., Sitompul, O. S., Nababan, E. B., Nababan, E. S. M., & Sinulingga, E. P. (2020). Data science around the indexed literature perspective. *Advances in Intelligent Systems and Computing*, 1294, 1051–1065. https://doi.org/10.1007/978-3-030-63322-6_91

15. Neji, S., Ben Ayed, L. J., Chenaina, T., & Shoeb, A. (2021). A novel conceptual weighting model for semantic information retrieval. *Information Sciences Letters*, 10(1), 121–130. https://doi.org/10.18576/isl/100114
16. Spink, A., & Saracevic, T. (1997). Interaction in information retrieval: Selection and effectiveness of search terms. *Journal of the American Society for Information Science*, 48(8), 741–761. https://doi.org/10.1002/(SICI)1097-4571(199708)48:8<741::AID-ASI7>3.0.CO;2-S
17. Blair, D. C. (2003). Information retrieval and the philosophy of language. *Annual Review of Information Science and Technology*, 37(3). https://doi.org/10.1002/aris.1440370102
18. Mitra, M. (2000). Information retrieval from documents: A survey. *Information Retrieval*, 2(2–3), 141–163. https://doi.org/10.1023/a:1009950525500
19. Arslan, A. (2016). DeASCIIfication approach to handle diacritics in Turkish information retrieval. *Information Processing and Management*, 52(2), 326–339. https://doi.org/10.1016/j.ipm.2015.08.004
20. Sharma, A., & Kumar, S. (2020). Semantic web-based information retrieval models: A systematic survey. *Communications in Computer and Information Science*, 1230 CCIS, 204–222. https://doi.org/10.1007/978-981-15-5830-6_18
21. Shen, Y., Chen, C., Dai, Y., Cai, J., & Chen, L. (2021). A hybrid model combining formulae with keywords for mathematical information retrieval. *International Journal of Software Engineering and Knowledge Engineering*, 31(11–12), 1583–1602. https://doi.org/10.1142/S0218194021400131
22. Mahalakshmi, P., & Fatima, N. S. (2022). Summarization of text and image captioning in information retrieval using deep learning techniques. *IEEE Access*, 10, 18289–18297. https://doi.org/10.1109/ACCESS.2022.3150414
23. Nasution, M. K. M. (2012). Simple search engine model: Adaptive properties. *Cornell University Library*. arXiv:1212.3906 [cs.IR]
24. Gascón, A., Godoy, G., & Schmidt-Schauş, M. (2009). Unification with singleton tree grammars. *Lecture Notes in Computer Science (including subseries Lecture Notes in Artificial Intelligence and Lecture Notes in Bioinformatics)*, 5595 LNCS, 365–379. https://doi.org/10.1007/978-3-642-02348-4_26.
25. Nasution, M. K. M. (2012). Simple search engine model: Adaptive properties for doubleton. *Cornell University Library*. arXiv:1212.4702 [cs.IR]
26. Lahiri, I. (2021). An entire function weakly sharing a doubleton with its derivative. *Computational Methods and Function Theory*, 21(3), 379–397. https://doi.org/10.1007/s40315-020-00355-4
27. Al-Kharashi, I. A., & Evens, M. W. (1994). Comparing words, stems, and roots as index terms in an Arabic information retrieval system. *Journal of the American Society for Information Science*, 45(8), 548–560. https://doi.org/10.1002/(SICI)1097-4571(199409)45:8<548::AID-ASI3>3.0.CO;2-X
28. Hines, D. (1979). Lilliputian library catalogs and information retrieval. *Journal of Micrographics*, 12(6), 317–322.
29. Krovetz, R., & Croft, W. B. (1992). Lexical ambiguity and information retrieval. *ACM Transactions on Information Systems (TOIS)*, 10(2), 115–141. https://doi.org/10.1145/146802.146810
30. Baker, F. B. (1962). Information retrieval based upon latent class analysis. *Journal of the ACM (JACM)*, 9(4), 512–521. https://doi.org/10.1145/321138.321148
31. Levine, G. R. (1981). Developing databases for online information retrieval. *Online Information Review*, 5(2), 109–120.
32. Alexandrov, V. N., Dimov, I. T., Karaivanova, A., & Tan, C. J. K. (2003). Parallel Monte Carlo algorithms for information retrieval. *Mathematics and Computers in Simulation*, 62(3–6), 289–295. https://doi.org/10.1016/S0378-4754(02)00252-5

33. Dapolito, F., Barker, D., & Wiant, J. (1971). Context in semantic information retrieval. *Psychonomic Science*, 24(4), 180–182. https://doi.org/10.3758/BF03335558
34. Quan, D. (2007). Improving life sciences information retrieval using semantic web technology. *Briefings in Bioinformatics*, 8(3), 172–182. https://doi.org/10.1093/bib/bbm016
35. Lee, K.-C., Hsieh, C.-H., Wei, L.-J., Mao, C.-H., Dai, J.-H., & Kuang, Y.-T. (2017). Sec-Buzzer: Cyber security emerging topic mining with open threat intelligence retrieval and timeline event annotation. *Soft Computing*, 21(11), 2883–2896. https://doi.org/10.1007/s00500-016-2265-0
36. Syahputra, I., Ritonga, R., Purwani, D. A., Masduki, Rahmaniah, S. E., & Wahid, U. (2021). Pandemic politics and communication crisis: How social media buzzers impaired the lockdown aspiration in Indonesia. *SEARCH Journal of Media and Communication Research*, 13(1), 31–46.
37. Bedi, P., & Chawla, S. (2007). Improving information retrieval precision using query log mining and information scent. *Information Technology Journal*, 6(4), 584–588. https://doi.org/10.3923/itj.2007.584.588
38. Zillmann, H. (2000). Information retrieval and search engines in full-text databases. *LIBER Quarterly*, 10(3), 334–341. https://doi.org/10.18352/lq.7605
39. Nasution, M. K. M. (2017). Modelling and simulation of search engine. *Journal of Physics: Conference Series*, 801. https://doi.org/10.1088/1742-6596/801/1/012078
40. Meštrović, A., & Calì, A. (2017). An ontology-based approach to information retrieval. *Lecture Notes in Computer Science (including subseries Lecture Notes in Artificial Intelligence and Lecture Notes in Bioinformatics)*, 10151 LNCS, 150–156. https://doi.org/10.1007/978-3-319-53640-8_13
41. Stokoe, C. (2005). Differentiating homonymy and polysemy in information retrieval. In *HLT/EMNLP 2005 — Human language technology conference and conference on empirical methods in natural language processing, proceedings of the conference*, pp. 403–410. https://doi.org/10.3115/1220575.1220626
42. Safi, H., Jaoua, M., & Belguith, L. H. (2016). PIRAT: A personalized information retrieval system in Arabic texts based on a hybrid representation of a user profile. *Lecture Notes in Computer Science (including subseries Lecture Notes in Artificial Intelligence and Lecture Notes in Bioinformatics)*, 9612, 326–334. https://doi.org/10.1007/978-3-319-41754-7_31
43. Bell, C. J. (1968). Implicit information retrieval. *Information Storage and Retrieval*, 4(2), 139–160. https://doi.org/10.1016/0020-0271(68)90017-X
44. Papadimitriou, C. H. (1998). Algorithmic approaches to information retrieval and data mining. *Lecture Notes in Computer Science (including subseries Lecture Notes in Artificial Intelligence and Lecture Notes in Bioinformatics)*, 1449, 1.
45. Nasution, M. K. M., & Noah, S. A. (2012). Information retrieval model: A social network extraction perspective. In *Proceedings—2012 international conference on information retrieval and knowledge management*. CAMP'12, pp. 322–326. https://doi.org/10.1109/InfRKM.2012.6204999
46. Nasution, M. K. M. (2018). Singleton: A role of the search engine to reveal the existence of something in information space. *IOP Conference Series: Materials Science and Engineering*, 420(1). https://doi.org/10.1088/1757-899X/420/1/012137
47. Nasution, M. K. M. (2018). Doubleton: A role of the search engine to reveal the existence of relation in information space. *IOP Conference Series: Materials Science and Engineering*, 420(1). https://doi.org/10.1088/1757-899X/420/1/012138
48. Xu, M., & Ke, Y. (2014). The research of information retrieval technology based on semantic analysis. *Advanced Materials Research*, 926–930, 2160–2163. https://doi.org/10.4028/www.scientific.net/AMR.926-930.2160

49. Hoque, M. N., Islam, R., & Karim, M. S. (2019). Information retrieval system in Bangla document ranking using latent semantic indexing. In *1st international conference on advances in science, engineering and robotics technology 2019*. ICASERT. https://doi.org/10.1109/ICASERT.2019.8934837
50. Kumar, R. (2022). Smart information retrieval using query transformation based on ontology and semantic-association. *International Journal of Advanced Computer Science and Applications*, 13(3), 388–394. https://doi.org/10.14569/IJACSA.2022.0130446
51. Zhumabay, R., Kalman, G., Sambetbayeva, M., Yerimbetova, A., Ayapbergenova, A., & Bizhanova, A. (2022). Building a model for resolving referential relations in a multilingual system. *Eastern-European Journal of Enterprise Technologies*, 2(2–116), 27–35. https://doi.org/10.15587/1729-4061.2022.255786
52. Zhang, J., Tan, X., Huang, X., & Wang, Y. (2009). Using semantic web for information retrieval based on clonal selection strategy. *ISCID 2009–2009 International Symposium on Computational Intelligence and Design*, 1, 513–516. https://doi.org/10.1109/ISCID.2009.135
53. Bassil, Y., & Semaan, P. (2012). Semantic-sensitive web information retrieval model for HTML documents. *European Journal of Scientific Research*, 69(4), 550–559.
54. Do, T. T.-T., & Nguyen, D. T. (2021). A computational semantic information retrieval model for Vietnamese texts. *International Journal of Computational Science and Engineering*, 24(3), 301–311. https://doi.org/10.1504/IJCSE.2021.115657
55. Mooers, C. N. (1954). Choice and coding in information retrieval systems. *IRE Professional Group on Information Theory*, 4(4), 112–118.
56. Cochran, R. (1959). Information retrieval study. *Proceedings of the Western Joint Computer Conference, IRE-AIEE-ACM 1959*, 283–285. https://doi.org/10.1145/1457838.1457891
57. Wolfe, T. (1971). Suggestions for exploiting the potential of on-line remote access information retrieval and display systems. *Journal of the American Society for Information Science*, 22(3), 149–152. https://doi.org/10.1002/asi.4630220302, https://asistdl.onlinelibrary.wiley.com/doi/10.1002/asi.4630220302
58. Negoita, C. V. (1973). On the notion of relevance in information retrieval. *Kybernetes*, 2(3), 161–165. https://doi.org/10.1108/eb005334
59. Wiersba, R. K. (1979). The role of information retrieval in the second computer revolution. *ACM SIGIR Forum*, 14(2), 52–58. https://doi.org/10.1145/1013232.511714
60. Bookstein, A. (1983). Information retrieval: A sequential learning process. *Journal of the American Society for Information Science*, 34(5), 331–342. https://doi.org/10.1002/asi.4630340504
61. Miyamoto, S., Miyake, T., & Nakayama, K. (1983). Generation of a pseudo thesaurus for information retrieval based on co-occurrences and fuzzy set operations. *IEEE Transaction on Systems, Man and Cybernetics, SMC*, 13(1), 62–70. https://doi.org/10.1109/TSMC.1983.6313030
62. Kraft, D. H. (1985). Advances in information retrieval: Where is that /#\&@¢ record? *Advances in Computers*, 24(C), 277–318. https://doi.org/10.1016/S0065-2458(08)60369-1
63. Gordon, M. D., & Fry, J. P. (1989). Novel application of information retrieval to the storage and management of computer models. *Information Processing and Management*, 25(6), 629–646. https://doi.org/10.1016/0306-4573(89)90097-6
64. Sharma, R. (1989). A generic machine for parallel information retrieval. *Information Processing and Management*, 25(3), 223–235. https://doi.org/10.1016/0306-4573(89)90041-1
65. Rau, L. F., & Jacobs, P. S. (1989). *NL* ∩ *IR*: Natural language for information retrieval. *International Journal of Intelligent Systems*, 4(3), 319–343. https://doi.org/10.1002/int.4550040306
66. Zhai, J., Cao, Y., & Chen, Y. (2008). Semantic information retrieval based on fuzzy ontology for intelligent transportation systems. In *Conference proceedings—IEEE international conference on systems, man and cybernetics*. CODEN PICYE, Singapore, pp. 2321–2326.

67. Wang, S.-D., Li, J.-H., & Wang, S.-N. (2011). Research on semantic information retrieval based on ontology. *Journal of Harbin Institute of Technology (New Series)*, 18(SUPPL. 1), 112–115.
68. Chauhan, R., Goudar, R., Rathore, R., Singh, P., & Rao, S. (2012). Ontology based automatic query expansion for semantic information retrieval in spots domain. *Communications in Computer and Information Science CCIS*, 305, 422–433.
69. Huang, Y., Li, G., & Li, Q. (2013). Rough ontology based semantic information retrieval. In *Proceedings—6th international symposium on computational intelligence and design, ISCID 2013*, 1, 63–67.
70. Remi, S., & Varghese, S. C. (2015). Domain ontology driven fuzzy semantic information retrieval. *Procedia Computer Science*, 46, 676–681.

5 Enterprise Application Development Using Semantic Web Languages for Semantic Data Modeling

D.V. Chandrashekar, Syed Md Fazal and K. Suneetha

CONTENTS

DOI: 10.1201/9781003310792-5

5.1 INTRODUCTION

Languages like Resource Description Framework (RDF), Resource Description Framework Schema (RDFS), and Web Ontology Language (OWL), which are all semantic web ontology languages, have been considered standards since their inception in the early 2000s.[1] Research on RDF and the semantic web, and on RDF/OWL data storage and query, is extensive. In order to ensure semantically oriented software development, this chapter recommends and encourages software engineers to use semantically oriented software development as a foundation. Our research [2] has been enriched as a result of this semantic premise. The relational database model as a foundation for traditional approaches to enterprise data modeling databases (RDBs) and RDB-oriented data modeling has been happening for many years, so almost all of these approaches are based on RDBs. Semantically oriented data modeling with RDF and OWL is becoming increasingly relevant in a variety of fields, despite the fact that the technologies are still maturing. Consider how frequently RDF is used to represent data and how RDF and OWL have both improved over time. As an added bonus, we're debuting and distributing a key model reference in the area of semantic information conception, retrieval, and manipulation for use in corporate applications. RDF and OWL are recommended for modeling data in software engineering, and we provide an example of how this can be done. A semantic data model is needed for data management and ontological applications, as we explain in this chapter. It is possible to make data more valuable by using RDF and OWL in the data architecture. Data interoperability, graph modeling, and model reuse, among other things, are handled by the technique. Semantic documentation and semantic validation are also included. A comparison between the semantic and relational approaches can also be made using this way.[3]

This chapter is organized as follows: Section 5.2 covers RDF, RDFS, and OWL ontologies. Section 5.3 focuses on data modeling research. Section 5.4 goes over semantic data modeling in great detail. Section 5.5 introduces our RDF/RDFS and OWL-based data modeling approach. Our methods are compared to those used in other relevant studies in Section 5.6, which deals with many of the same topics. Finally, Section 5.7 concludes this chapter. Everything is covered here.

5.2 RDF, RDFS, AND OWL

The World Wide Web Consortium (W3C) promotes RDF, RDFS, and OWL as primary languages for semantic web implementation.[3–5]

5.2.1 RDFS AND RDF

In RDF, triples are used to represent resources and their relationships (S, P, and O). As a type of resource, "S" stands for "subject." There are two ways to define the predicate p: as a resource or as an abstract representation of the subject-object relationship. There are numerous methods for serializing RDF; N-Triples (notation 3), Turtle (notation 4), XML (notation 5), and JSON are the most widely used formats. RDF uses sequences and lists to express ordered lists of resources or literal values; the type "Bag" can represent collections of resources or literal values that are not organized in any particular way.

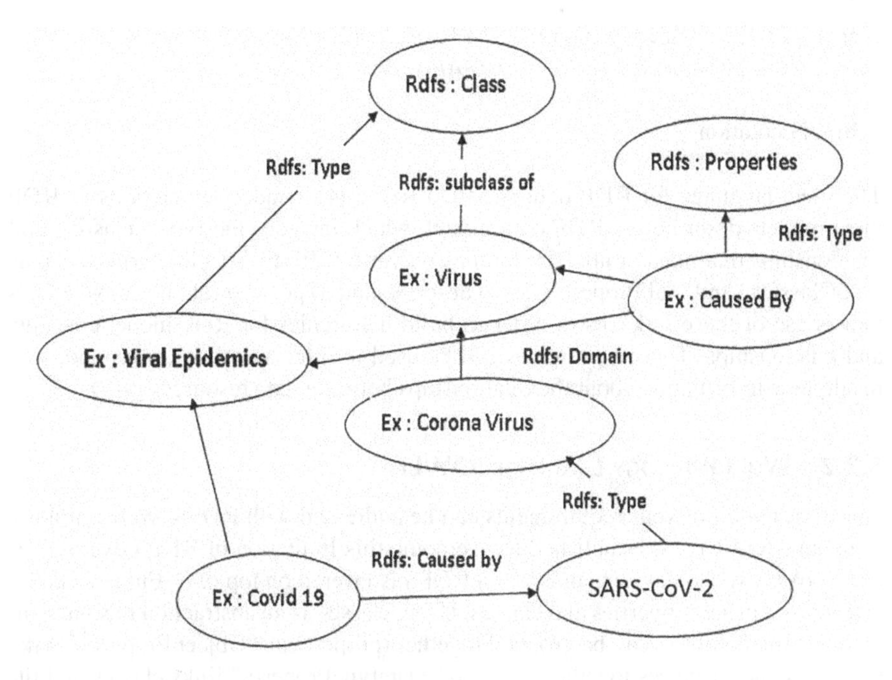

FIGURE 5.1 Example of RDF/RDFS triples.

Figure 5.1 depicts an RDF/RDFS example of SARS-CoV-2 and the epidemic "Covid-19," as well as the classes connected with them. For example, we have the triples (ex: virus, rdfs: subClassOf, rdfs: class), (ex: viral Epidemic, rdf: type, rdf: class), (ex: caused By, rdf: domain, ex: virus), and (ex: caused By, rdf: domain, ex: virus) (ex: covid-19, ex: caused By, ex: SARS-CoV-2).

```
ex:virus rdfs:subClassOf rdfs:Class.
ex:viralEpidemic rdf:type rdf:class.
ex:causedBy rdf:type rdfs:Property.
ex:causedBy rdf:domain ex:viralEpidemic.
ex:causedBy rdf:range ex:virus.
ex:coronavirus rdfs:subClassOf ex:virus.
ex: SARS-CoV-2 rdf:type ex:coronavirus.
```

(a) Graphical presentation

```
ex:virus rdfs:subClassOf rdfs:Class.
ex:viralEpidemic rdf:type rdf:class.
ex:causedBy rdf:type rdfs:Property.
ex:causedBy rdf:domain ex:viralEpidemic.
ex:causedBy rdf:range ex:virus.
ex:coronavirus rdfs:subClassOf ex:virus.
ex: SARS-CoV-2 rdf:type ex:coronavirus.
```

```
ex:covid-19 rdf:type ex:viralEpidemic.
ex:covid-19 ex:causedBy ex:SARS-CoV-2.
```

(b) N3 notation

The meta-language for RDF data is called RDFS.[4] Triadic descriptions of RDF meta-models describe the RDF data model, which includes the types, classes, and relationships that make it up. RDFS relies on various RDFS-specific terms, such as "subClassOf" and "subPropertyOf," to describe data types and relationships. RDFS makes use of these linkages in order to build hierarchies for RDF model concepts and relationships. These hierarchies can be used to infer new data triples and, as a result, new information about the relationships between data resources.[5–7]

5.2.2 Web Ontology Language (OWL)

However, not all of your requirements can be addressed with RDFS. Web Ontology Language (OWL) was established to overcome this limitation of RDFS expressiveness. OWL's vocabulary expands when RDFS is layered on top of it. Entities can be categorized using properties and classes. Using classes as an abstraction mechanism, similar functionalities can be grouped together (properties). "Object Property" links resources and instances together, whereas "DatatypeProperty" links classes and literal values.

Several range and domain constructs, such as SomeValues From, ComplementOf, AllValues From, and DisjointWith are accessible in OWL. Inverse functional qualities include things like symmetry, inverse, transitivity, and function. It is possible to create new data types from existing ones using OWL components such as "DataUnionOf," "DataComplementOf," or a "DataIntersectionOf.".

5.2.3 SPARQL is a Query Language

SPARQL Protocol and RDF Query Language is a recursive acronym. The RDF query language is represented in this language. SPARQL is an official W3C recommendation for extracting data from RDF. Select, construct, ask, and describe are SPARQL's four query types (as in, "tell me about yourself"). When a query pattern match is detected, the select form delivers a list of variables and their bindings. RDF graphs generated by the construct form are the result of your input. In the ask form, you can check to see if a question has an answer available or not. An RDF graph summarizing the resources found by the describe form is provided. It's SPARQL Complex RDF joins can also be done in one query, which results in the results being integrated from multiple RDF sources.[8]

5.3 COLLABORATIVE RESEARCH

RDF/OWL models (ontologies [9–13]) have been developed from relational models (such as UML) by researchers. Research on ontology frameworks has focused on how to create and publish ontologies for certain subjects on the semantic web without taking into account how enterprise applications' data models are structured.[14] Researchers looked at the groundwork that has been laid in this area. Data migration to the semantic web is discussed from a commercial standpoint in Alaoui et al. (2014).[15] Ontological data from the semantic web can be reverse mapped to relational models in other publications.[16–18] Benslimane et al. (2010) deals with ontological modeling, but it only discussed how ontology can aid in the development of model objects.[19] However, ontologies have not been used to model business application data in software engineering, only to model the control of activities during the development process. A number of previous studies in this area have tackled the issue of software engineering task traceability by utilizing control ontologies to specify process tasks or traceability difficulties connected to the creation of documents and decisions.

5.4 DATA MODELING BASED ON SEMANTICS

Relational databases (RDBs) have increased in popularity for a number of years, and, as a result, data models for applications have gotten increasingly relational. The various domain elements are described through entities and relations with attributes. Due to its long-term development and general availability, relational database management systems (RDBMS) became widely used and were a key factor in handling data more efficiently. Relational data modeling is being encouraged by new tools and APIs for interfacing with these systems.[20] A semantic approach to data modeling based on RDF and OWL will be explained in the following sections.

5.4.1 Acceptance of RDF

Over the past few years, RDF and OWL have become increasingly popular in a variety of industries (e.g., geology [21]). Biomedicine, government data, academic social networks, DBpedia, and tourism are all included under the term "smart cities," which encompasses a wide range of subjects.

5.4.2 RDF Management Systems Come in All Shapes and Sizes

Data management technologies, such as triplestores, have arisen as a result of this widespread usage. They have been built with the inclusion of further tools and APIs for RDF/OWL data presentation and interoperability with other triplestores. For further information about RDF triple store categories, please consult Challenger (2012).[22] General-purpose triplestores, NoSQL triplestores, P2P triplestores, and cloud triplestores are only some of the current triplestores available. Only a few examples are given in Chawla et al. (2019) and Cossu et al. (2018).[23, 24] The types of data that can be stored in triplestores range from relational to native and from centralized triplestores to distributed triplestores (e.g., Triple bit [25]). Trinity RDF,[26] for examplc, or Triad,[27] are good examples of this. Many RDF stores have been developed to handle large amounts of RDF data, including big data RDF stores. These big data RDF stores rely on Hadoop Distributed File System (HDFS) and the MapReduce framework for SPARQL queries on RDF triples. It is straightforward and concise in Gurajada et al. (2014) how the MapReduce meta-model works.[28]

SPARQLGX,[29, 30] S2RDF,[31] S2RDF-Spark,[32] and PRoST,[33] as well as PRESTO-RDF,[34] SHARD,[35] HadoopRDF,[36] and CliqueSquare,[37] are just a few instances of this type of solution. RDF management systems from big data and NoSQL systems are compared in Nalepa and Furmanska (2009).[38] The development of RDF big data triplestores has also been supported by the expansion of a wide range of analytical tools and technologies in recent years. Rajbhandari et al. (2012) is a good place to start learning about these new tools and technology.[39]

Data can be modeled using RDF/RDFS and OWL utilizing any of the previously-mentioned RDF/OWL management systems.

5.4.3 Data Modeling using RDF Data Tools

1. Information based on the Resource Description Framework
2. Changes have been made to the tools for editing and exploring RDF schemas along with the evolution of semantic web standards.
3. When dealing with large amounts of data, these technologies are crucial.
4. RDF/RDFS data visualization is, of course, essential. Because of this visual representation of data, end users are better able to grasp and deal with the models they are provided with. A tool that we utilize is D3SPARQL.
5. RDF Instance Creator, Protégé, and Onto Edit (RIC).
6. RDF Instances (RIC): RDF Instances (RIC) are created, promoted, and edited.
7. NoSQL-based triplestores, for example, provide their own visualization APIs as well.
8. A set of tools for working with RDF.
9. Data modeling techniques are typically used to build systems and APIs that help users organize and manage their operations. So many triplestores have been created in this fashion with an API for working with RDF data and schemas to their credit.
10. The most well-known and widely utilized is Jena. You can generate and manipulate RDF graphs in Java by utilizing the Jena RDF API. This application, which works with XML and N-Triples RDF files, displays RDF graphs as "models."
11. In addition to Eclispe-rdf4j, another Java-based RDF data API enables SPARQL queries. RDF4J is capable of reading and writing RDF files in XML, Turtle, and N-Triples.

There are APIs for a variety of programming languages and frameworks, as well. The ActiveRDF Ruby library provides access to RDF data, for example, due to the broad usage of RDF and OWL in numerous domains, and the advances achieved in storing, presenting, and manipulating RDF and OWL data and semantic data modeling should not be limited to disseminating such data as the semantic web intended. Being truly valuable, knowledge must be rich enough in semantics to be conveyed by machines, by users, and by applications themselves. In a group, the same ontological sets can be used by humans and machines without the need for adjustments. With semantic navigation, different organizations can easily use their own data models and get to semantic information.[40]

Helping the numerous RDF and OWL user communities who have adopted or plan to use RDF and OWL in a way that respects their semantic goals in the development of software they need is crucial. If we want to avoid all of the terrible repercussions that followed the software crisis, we need to avoid the issues brought to light by that crisis. For software that may communicate semantically with its users, there are additional issues for the semantics of software development, as well as all of the other hurdles of the standard software development process.[41, 42] For semantically oriented data modeling, we propose a five-step approach. In this section, we lay out a framework for implementing the semantic data modeling approach we've advocated.

5.5 ACTIVE OPERATIONAL COMMAND BASED ON SEMANTICS

There must be intimate ties and consideration of the demands of the target audience in these models in order for them to engage with them. Semantic data modeling necessitates this step. There should be a common understanding of what a model is saying among both users and machines alike. The following lists the most common approaches to semantic-oriented analysis and conception. A set of semantics is discovered.[43]

In order to get meaning from data models modeled in RDF or OWL, semantics must be extracted. The project aims to develop semantic RDF data models. A specific use case domain defines the semantics of the various data resources and their linkages while working with RDFS and OWL in this manner. When you look at the ontology or graph that represents a definition, you can learn a lot about what the idea is really about. The emphasis here is on semantic data interoperability. Semantic data interoperability is another word for connecting semantically oriented data. A graph-based view of the data can be created since RDF and OWL can function together. Graph theory is a field of study in academia. By using RDF as a model for triples, it is possible to manage and retrieve information as well as easily add new components to a graph model. Cross-referencing and mislabeling of data are to blame. Additional graph operations, including path search, graph isomorphism, graph matching, and connectivity testing, enable graph exploration for semantic navigation and information extraction to more effectively avoid the re-use of the same model.[44, 45] It is possible to repurpose existing models. As a result, anthologies can be used by a wide range of software applications and procedures. Reasoning with ontologies and models, simple logic can be used to generate information from models like RDFS and OWL, which can then be represented using models like RDFS and OWL. Because implicit semantics can be inferred from existent semantics, redundancy is avoided thanks to the capacity to reason over models. Thus, new triples may be discovered by applying logic to previously discovered pairs.[46, 47]

Data models and RDF schemas can be worked with using readily available tools for analytical research editors. XML or graphing tools can be used to alter RDF. Every time a user edits an XML document, they must confirm that the elements' syntax adheres to the standard. They could be enhanced with RDF-based verification.

Graph visualization tools and editors make it feasible to compare data models and find problems via graphical representation. Graphical tools provide greater control over the RDF schema's content, which can be rather vast and spread out across numerous files. The structure and interconnections of each piece can be better understood through the use of visual and diagrammatic presentations. It is possible to create schemas from the reworked graphic models. Communication among project participants is facilitated through the use of these visual representations.[48, 49]

5.5.1 Documentation that Focuses on Semantics

Semantically oriented documentation demands an explanation of the reasoning behind the evaluation of each graph or subgraph model. It's imperative that users and linked domains or subdomains are protected from this. To make sure that data models are consistent and that users are happy, you must use user interactions and inference tests.

5.5.2 Elasticity and Multi-Modeling

A robust data model should be able to deal with enormous amounts of data as well as processing that is both simple and sophisticated.

5.5.3 Integrated Semantic Data

In today's world, you'll need to be able to deal with a plethora of data originating from numerous sources. It's critical that diverse apps, such as cloud, mobile, and big data ones, can communicate and share data in order to get the most out of them. There should be a comprehensive framework for integrating semantic data into a model's overall theory. The RDF and OWL models are ideal for integrating data from a variety of sources that are distinct from one another. When you mix data from many sources, you are forced to consider a wide range of considerations. A few examples include dealing with the words used and finding synonyms for those you currently possess. This could lead to the creation of new graphs or subgraphs, or it could simply lead to the addition of new data to existing graphs. Because we have a variety of data sources, such as relational, XML, and NoSQL, we must deal with a variety of data kinds and formats. Another consideration is the wide range of data. Decisions must be made within the context of meaning in order for the new models to be added to the global ones. In order to deal with the data's structure and properties, it's best to have a conversion platform that can directly extract the RDFS schema as well as any associated data graphs. This might be accomplished using existing mapping and conversion tools and methodologies.[50]

5.6 TRADITIONAL RDB-ORIENTED DATA MODELING HAS LESS IMPACT

This section compares RDF and OWL-based semantic data modeling combined with traditional data modeling techniques outlines the advantages of each.

5.6.1 RDB-Oriented Data Modeling in the Past Semantic Modeling on the Cloud (DSM)

The graph modeling approach excels at scalability. The graph's ability to divide makes it easy to see the distribution of data. Multi-modeling can also be used to create linked schema models, cloud, and peer-to-peer network modeling of massive volumes of data. Data generated by RDF and OWL can be easily maintained by networked information management systems in peer-to-peer or cloud applications. Several cloud or P2P data management systems are discussed in Zeng et al. (2013) and Zhang et al. (2011).[51, 52] When dealing with massive amounts of data, big data technologies like Hadoop and Spark can be used to deploy triplestores. Classical data modeling focuses on relational databases. To make a conceptual data model, this model breaks down the requirements into their parts. This lets the entities, their properties, and the connections between them be found. Based on the database system, the conceptual data model is then used to create a logical data model. This physical data model can then be used. Because of these sequences of changes, relational database-oriented modeling is not suitable for semantic-oriented modeling. Words can lose some of their original meaning when they are rephrased. Due to the rigidity of relational model rules, the significance of the data may be lost. As a result, RDB data models fall short of the flexibility and quality required for web semantics.

5.6.2 Advantages to Semantic Data Modeling

Semantic data modeling relies heavily on graph-oriented modeling based on RDF and OWL. RDF and OWL-based data modeling is more expressive and flexible than relational data modeling. By defining the semantics of the pieces they want to model, users can quickly and easily construct or add models. This can be done without any restrictions or modifications. All user requirements can be accurately recorded using RDF and OWL-based models, which focus more on the nature of the data. In addition, this modeling art does not depend on how it's used, so there are no technological worries or revisions required. Furthermore, it is simple to expand and compare the models.[53]

RDF, RDFS, and OWL are strong data modeling technologies with benefits. It is from the data model that developers, as well as customers and end users, may begin their discussions. Since this is a tool that is used frequently, it should be as plain and straightforward as possible. Thus, the RDF modeling technique provided here better meets these requirements because it focuses on discovering the semantics of real-world items and then expressing them as RDF resources with characteristics that illustrate how these entities are connected in the form of navigable graphs. This strategy has the advantages listed below:

- RDF/capacity of OWLs to build clear semantic formulations where the meaning can be determined from the sequence in which they are expressed (S, P, O).
- Easy navigation of models thanks to the readability of RDF/RDFS graphs. Fast and simple data modeling utilizing (subject, predicate, and object) triples.

- SPARQL can search and query schema models since they are written in RDF as (S, P, O) triples.
- Conceptual models are logical models that do not repeat existing schema triples. So, they can be uploaded directly to the vast majority of existing RDF triplestores.
- No intermediate models or database designers are required.
- Data models that mirror the way users think, such as RDF and OWL semantic-oriented models, can lead to innovative enterprise applications and services.
- Models can be reused and distributed via graph-based semantic modeling. Graphs are also more expressive, and they allow for more efficient retrieval of data. For example, there are many algorithms for graph operations like matching and isomorphism that are used today.
- RDF Schema constructs can be easily browsed using existing RDF tools.
- RDF graphs can be used to communicate with end-users at any time.
- Serialization and interoperability.
- RDF models and data can also be used to serialize RDF models and data, making them interoperable with a wide range of existing tools and languages.
- Methods for the conversion or mapping of numerous data sources have been developed in recent years, including Extensible Markup Language (XML) and Unified Modeling Language (UML), as well as relational data sources. To add a new data source to an existing system, semantics should be taken into consideration.
- RDF data and schemas can be interacted with in a variety of ways using programming languages like Java, Perl, and others. Because E/R-modeling has a limited ability to utilize these qualities, they are extremely crucial.

When data is linked in models, the semantic value of each data property and component goes up. Semantic modeling is an effective way to find information because there are many graph algorithms that can be used to do things like graph matching and graph isomorphism.

The RDF framework can be used to process data from P2P, cloud, and big data architectures respectively. You are able to acquire a deeper comprehension of the systems at hand by using models that are based on research from the past and actions that can be taken based on what you know about the domain and what users expect.

5.7 COMPARISON OF SEMANTIC WEB SERVICE APPROACHES

The semantic web is an extension of the traditional web that makes it easier for people and machines to find information, distribute it, reuse it, and integrate it into other systems. It all boils down to an ontology providing the semantics for a web, which is essentially what it is used for. This architecture allows for the publication of an ontological explanation of terms and concepts, as well as any conflicts that may arise as a result of their use. As a direct consequence of this, semantic interoperability of web services is made attainable by the identification (and mapping) of concepts that are semantically connected to one another. Techniques such as resolution and mapping between anthologies are included in this category.[53] Similar to regular web services, semantic web services have the potential to be improved through the application of semantic annotations in order to better characterize the services. This enables the discovery of services, the binding of those services, and the invocation of those services to be carried out automatically. A further advantage of utilizing SWS is that it can facilitate the seamless integration of a number of resources and services inside the environment of a corporation. Implementing semantic web services with WSMO, OWL-S, and IRS-III are three of the most promising approaches now available.

Ontological requirements that are known as the Web Service Modeling Ontology (WSMO) are meant to characterize the several components that make up semantic web services (SWS). The ultimate goal of this project is to create an integrated technology that will usher in the next generation of the Internet, shifting its focus from being a repository of information for human consumption to a global system for web-based distributed computing. Ontologies, web services, objectives, and mediators can all be found at the highest pinnacle of the pyramid. Ontology-based semantics offer a formal foundation for this terminology, which is used by all of the other WSMO components (which all employ the language specified by anthologies). If they so want, service providers have the option of using web service to communicate with their customers about the capabilities and procedures associated with the services they provide. In addition to that, it describes how web services communicate with one another and how they are choreographed (coordination).[53]

Service requesters will use goals to determine the qualities and demeanors that must be present in a service in order for it to fulfill all of their needs. Ontologies can be broken down into two distinct categories: those that specify semantically defined terminology and those that reflect user viewpoints inside SWS architecture (the latter of which is referred to as a user view). Mediators are responsible for resolving any data or protocol incompatibilities that may emerge between the components that are supposed to function together. These incompatibilities may arise due to the fact that different components use different protocols.

A recently developed ontology is known as OWL-S, which stands for "Ontology for the Description of Semantic Web Services Represented in OWL." In order to describe services that can be articulated semantically while still remaining firmly grounded in a well-defined data type formalism,[54] the pragmatism of the developing web services standards is combined with the expressivity of description logics. This combination is called "description logics." For instance, because of this, it is able to combine the expressiveness of description logics with the pragmatism that will be present in the forthcoming web services standards. These are the three fundamental ideas: the service profile, the service model, and the service grounding. Each of these three ideas is tied to the others.

1. For the purpose of service discovery, a service profile is a representation of both the functional and non-functional properties of a web service. These features can be categorized as either public or private.
2. A description of the steps involved in putting together a service is a good illustration of the kind of service that the service model can provide as an

example. This can be utilized for the purpose of regulating the posting and invocation of services as well as conceiving of possible compositions.

3. Service grounding transforms an abstract specification into a concrete one by providing a description of how the service should be used.

5.7.1 Internet Reasoning Service

The Internet Reasoning Service, also known as IRS-III, is a tool that can be utilized to facilitate the creation and deployment of semantic web services. Utilizing this technology to construct semantically enhanced systems, either in their entirety or in part, can be done so through the use of the Internet. IRS-III is distinguished from past work on semantic web services by four key characteristics that were incorporated into its design.

1. Code written in a programming language, such as Java or Lisp (both of which are currently supported), can be automatically transformed into a web service by having the proper wrapper for that code automatically generated. Existing software has the potential to rapidly be converted into web services by utilizing this feature.
2. The IRS-III is able to provide assistance for capability-driven service execution. Users of the IRS-III programmed will have the ability to directly access Internet services by setting goals within the application.
3. The IRS-III can be designed to carry out a wide variety of functions in a number of different ways. Several essential components of IRS III can be switched out for the users' own custom-built semantic web services.
4. The IRS-III's are compatible with online services and can be utilized with them. To put it another way, every IRS-III service has the potential to be published as a traditional web service, and when interacting with other web service infrastructures, it will appear as though it were published in this manner.

Underpinnings such as WSDL and Simple Object Access Protocol (SOAP) are utilized as the foundation for both WSMO and OWL-S. OWL-S does not provide an explicit definition of choreography,[55] as opposed to WSMO's service interface description, which places an emphasis on a process-based explanation of how complex web services summon atomic web services. In WSMO's service interface description, this is not the case. WSMO offers a service interface description that encompasses orchestration and choreography rather than offering a specification of either one separately. The IRS-III ontology also makes use of the WSMO ontology. Developers of semantic web services have focused their efforts on simplifying the process by which users can quickly transform their existing service code into a semantic web service. This process can be completed in a matter of minutes. There have been many research and efforts done to improve and semanticist applications in many aspects of the enterprise web, such as decision support e-commerce,[56] supply chain management, resource discovery, and so on. Some of the few aspects include in the area of E-commerce, activity based on geographical area, M-commerce, supply chain and much more. This area encompasses a wide range of activities, including electronic commerce, management of supply chains, and more. Additionally, there have been a few attempts made to design new architectural frameworks that make it possible to integrate different business applications.[57–60]

5.8 CONCLUSION

This chapter discusses the use of RDF, RDFS, and OWL in enterprise application development using semantically oriented software engineering data modeling. Until now, no one has proposed a proposal as ground-breaking as this one. This method is better than the usual entity-relationship modeling approach when it comes to modeling different kinds of data from a variety of services that we use on a regular basis. An approach to semantic data architecture has been devised that focuses on the most effective methods. For data interoperability, graph modeling, model reuse and reasoning, this method is the best option.

Semantics, scalability, and several models are all forms of documentation and validation that are required. Moreover, we proved why rich semantic data models may be built using RDF/RDFS and OWL-based data modeling than relational data modeling does as well. It is possible that our method, which treats data in a more meaningful manner, will assist data scientists and enterprise software developers. The semantic web will be more widely adopted if RDF and OWL-based applications are used instead of RDF and OWL conversions. Semantically-oriented development processes, from conceptualization to delivery and maintenance, might benefit greatly from ongoing work to establish an all-encompassing methodology that incorporates RDF and OWL. Future initiatives will adopt this global strategy.

REFERENCES

1. URL: http://linkedgeodata.org/
2. URL: http://reference.data.gov.uk/
3. URL: www.w3.org/RDF/
4. URL: www.w3.org/TR/rdf-schema/
5. URL: www.w3.org/TR/owl2-overview/
6. URL: www.w3.org/TR/rdf-sparql-query/
7. URL: www.w3.org/TR/rdf-sparql-query/#QueryForms
8. URL: www.w3.org/2001/sw/rdb2rdf/wiki/Implem entations
9. URL: http://wiki.dbpedia.org/services-resources/ontology
10. Zhang, W., Paudel, B., Zhang, W., Bernstein, A., & Chen, H. (2019). Interaction embeddings for prediction and explanation in knowledge graphs. In *Proceedings of the twelfth ACM international conference on web search and data mining. WSDM '19*. New York, NY: ACM, pp. 96–104.
11. Abadi, D., Marcus, A., Madden, S., & Hollenbach, K. (2009). SW-store: A vertically partitioned DBMS for Semantic Web data management. *VLDB Journal*, 18(2). https://doi.org/10.1007/s00778-008-0125-y
12. Afzal, Z. H., Waqas, M., & Naz, T. (2016). OWLMap: Fully automatic mapping of ontology into relational database schema. *International Journal of Advanced Computer Science and Applications*, 7(11), 7–15. https://doi.org/10.14569/IJACSA.2016.071102

13. Alaoui, K., & Bahaj, M. (2020). Semantic oriented data modeling based on RDF, RDFS and OWL. In M. Ezzyani (Ed.), *Proceedings. Advanced intelligent systems for sustainable development (AI2SD'2019), volume 4 — advanced intelligent systems for applied computing sciences*, vol. 1105. Springer AISC, pp. 411–421. https://doi.org/10.1007/978-3-030-36674-2_42
14. Alaoui, K. (2019, October 2–4). A categorization of RDF triple stores. In *Proceedings smart city applications, SCA-2019*. Casablanca, Morocco: ACM, ISBN 978-1-4503-6289-4/19/10. doi:10.1145/3368756.3369047
15. Alaoui, L., El Hajjamy, O., & Bahaj, M. (2014). RDB2OWL2: Schema and data conversion from RDB into OWL2. *International Journal of Engineering & Research Technology IJERT*, 3(11).
16. Bagui, S., & Bouressa, J. (2014). Mapping RDF and RDF-schema to the entity relationship model. *Journal of Emerging Trends in Computing and Information Sciences*, 5(12).
17. Banane, M., & Belangour, A. (2019, July). An evaluation and comparative study of massive RDF Data management approaches based on big data technologies. *International Journal of Emerging Trends in Engineering Research*, 7(7). https://doi.org/10.30534/ijeter/2019/03772019
18. Bellini, P., & Nesi, P. (2018). Performance assessment of RDF graph databases for smart city services. *Journal of Visual Languages and Computing*, 45, 24–38.
19. Benslimane, S., Malki, M., & Bouchiha, D. (2010). Deriving, conceptual schema from domain ontology: A web application reverse engineering approach. *The International Arab Journal of Information Technology*, 7(2), 167–176.
20. Bizer, C., Lehmann, J., Kobilarov, G., Auer, S., Becker, C., Cyganiak, R., & Hellmann, S. (2009). DBpedia—a crystallization point for the web of data. *Journal of Web Semantics*, 7(3), 154–165, 2009. https://doi.org/10.1016/j.websem.2009.07.002
21. Calero, C., Ruiz, F., & Piattini, M. (2006). *Ontologies for software engineering and software technology*. Spain: Springer Nature.
22. Challenger, M. (2012). The ontology and architecture for an academic social network. *International Journal of Computer Science*, 9(1), 22–27.
23. Chawla, T., Singh, G., Pilli, E. S., & Govil, M. C. (2019). Research issues in RDF management systems. In *Proceedings ACM India joint international conference on data science and management of data CoDS-COMAD'19*. ACM online Publishing House, Kolkata, India, pp. 188–194.
24. Cossu, M., Färber, M., & Lausen, G. (2018, March 26–29). PRoST: Distributed execution of SPARQL queries using mixed partitioning strategies. In *Proceedings 21st international conference on extending database technology (EDBT)*. IANG Publishing House, Hongkong.
25. Dhankhar, A., & Solanki, K. (2019, November). A comprehensive review of tools & techniques for big data analytics. *International Journal of Emerging Trends in Engineering Research*, 7(11). www.warse.org/IJETER/static/pdf/file/ijeter257 11 2019.pdf
26. El Hajjamy, O., Alaoui, K., Alaoui, L., & Bahaj, M. (2016). Mapping UML to OWL2 ontology. *Journal of Theoretical and Applied Information Technology*, 90(1).
27. Erraissi, A., & Belangour, A. (2019). Meta-modeling of big data management layer. *International Journal of Emerging Trends in Engineering Research*, 7(7), 36–43. https://doi.org/10.30534/ijeter/2019/01772019
28. Gurajada, S., Seufert, S., Miliaraki, I., & Theobald, M. (2014). Triad: A distributed shared-nothing rdf engine based on asynchronous message passing. In *Proceedings ACM SIGMOD*.
29. Goasdoué, F., Kaoudi, Z., Manolescu, I., Quiané-Ruiz, J., & Zampetakis, S. (2016). *Clique square: Efficient*. Cham: Springer, pp. 80–87. https://doi.org/10.1007/978-3-319-46547-0_9
30. Graux, D., Jachiet, L., Genevès, P., & Layaïda, N. (2016). SPARQLGX: Efficient distributed evaluation of SPARQL with Apache spark. In P. Groth et al. (Eds.), *Proceedings ISWC 2016*, vol. 9982. Springer LNCS Notes, Germany.

31. Happel, H. J., & Seedorf, S. (2006). Applications of ontologies in software engineering. In *Proceedings second international workshop semantic web enabled software Engineering*, HSG Press, Italy.
32. Husain, F., McGlothlin, J., Masud, M. M., Khan, L., & Thuraisingham, B. (2011, September). Heuristics-based query processing for large RDF graphs using cloud computing. *IEEE Transactions on Knowledge and Data Engineering*, 23(9), 1312–1327, IEEE Xplorer, Singapore.
33. Iqbal, R., Murad, M. A. A., Mustapha, A., & Sharef, N. M. (2013). An analysis of ontology engineering methodologies: A literature review. *Research Journal of Applied Sciences, Engineering and Technology*, 6.
34. Kaoudi, Z., & Manolescu, I. (2015). RDF in the clouds: A survey. *The VLDB Journal*, 24(1), 67–91.
35. Kochut, K., & Janik, M. (2007). SPARQLeR: Extended Sparql for semantic association discovery. In *Proceedings 4th European semantic web conference,* NNS BRUCK Publication, Innsbruck, Austria.
36. Mammo, M., Hassan, M., & Bansal, S. K. (2015). Distributed SPARQL querying over big RDF data using PRESTO-RDF. *International Journal of Big Data*, 2(3).
37. Naacke, H., Amann, B., & Curé, O. (2017). SPARQL graph pattern processing with Apache Spark. In *Proceedings fifth international workshop on graph data-management experiences and systems, GRADES 2017*. New York: ACM, pp. 1:1–1:7.
38. Nalepa, G. J., & Furmanska, W. T. (2009). Review of semantic web technologies for GIS. *Automatyka*, 13(2), 485–492.
39. Rajbhandari, J. P., Gosai, R., Shah, R. C., & Pramod, K. C. (2012). Semantic web in medical information systems. *The of Advances in Engineering & Technology*, 5(1), 536–543.
40. Ricca, F., Grasso, G., Liritano, S., Dimasi, A., Lelpa, S., Manna, M., & Leone, N. (2010). A logic-based system for e-Tourism. *Fundamenta Informaticae*, 105, 35–55.
41. Rizzo, G., Di Gregorio, F., Di Nunzio, P., Servetti, A., & De Martin, C. (2009, July). A peer-to-peer architecture for distributed and reliable RDF storage. In *Proceedings IEE 1st international conference on networked digital technologies*. New York: Ostrava, pp. 28–31, pp. 94–99. https://doi.org/10.1109/NDT.2009.5272090
42. Rohloff, K., & Schantz, R. E. (2010, October 17–21). High-performance, massively scalable distributed systems using the MapReduce software framework: The SHARD triple-store. In *Proceedings programming support innovations for emerging distributed applications*. Reno, NEVADA Publishing House, pp. 1–5.
43. Schätzle, A., Przyjaciel-Zablocki, M., Skilevic, S., & Lausen, G. (2016). S2RDF: RDF querying with SPARQL on spark. *VLDB*, 804–815. https://doi.org/10.14778/2977797.2977806
44. Spanos, D. E., Stavrou, P., & Mitrou, N. (20120. Bringing relational databases into the semantic web: A survey. *Semantic Web Journal*, 3(2), 169–209.
45. Saikaew, K. R., Asawamenakul, C., & Buranarach, M. (2014). Design and evaluation of a NoSQL database for storing and querying RDF data. *KKU Engineering Journal*, 41(4), 537–545.
46. Siricharoen, W. V. (2009). Ontology modeling and object modeling in software engineering. *International Journal of Software Engineering and its Applications*, 3(1).
47. Soualah-Alila, F., Faucher, C., Bertrand, F., Coustaty, M., & Doucet, A. (2015, October). Applying semantic web technologies for improving the visibility of tourism data. In *Proceedings Eighth workshop on exploiting semantic annotations in information retrieval ESAIR'15*. Melbourne, Australia: ACM, pp. 5–10.
48. Stutz, P., Verman, M., Fischer, L., & Bernstein, A. (2013, October 21–22). Triple rush: A fast and scalable triple store. In *Proceedings 9th international workshop on scalable semantic web knowledge base systems*. ACM Publishing, Sydney, Australia.
49. Vrandecic, D., & Krötzsch, M. (2014, September). Wikidata: A free collaborative knowledgebase. *Communications of the ACM*, 57(10), 78–85.

50. Yuan, P., Liu, P., Wu, B., Jin, H., Zhang, W., & Liu, L. (2013, May). Triplebit: A fast and compact system for large scale rdf data. *Proceedings VLDB Endowment*, 6(7), 517–528. https://doi.org/10.14778/2536349.2536352
51. Zeng, K., Yang, J., Wang, H., Shao, B., & Wang, Z. (2013). A distributed graph engine for web scale rdf data. In *Proceedings VLDB endowment*, vol. 6, Issue 4, pp. 265–276. https://doi.org/10.14778/2535570.2488333
52. Zhang, F., Ma, Z. M., & Yan, L. (2011). Construction of ontologies from object-oriented database models. *Integrated Computer-Aided Engineering*, 18, 327–347.
53. Zhao, Y., & Dong, J. (2009). Ontology classification for semantic-web-based software engineering. *IEEE Transactions on Services Computing*, 2(4). https://doi.org/10.1109/TSC.2009.20
54. Wan, S., Mak, M. W., & Kung, S. Y. (2016). Mem-mEN: Predicting multi-functional types of membrane proteins by interpretable elastic nets. *IEEE/ACM Transactions on Computational Biology and Bioinformatics*, 13(4), 706–718.
55. Wang, H., Zhang, F., Wang, J., Zhao, M., Li, W., Xie, X., & Guo, M. (2018). Ripplenet: Propagating user preferences on the knowledge graph for recommender systems. In *Proceedings of the 27th international conference on information and knowledge management*. CIKM '18. New York, NY: ACM, pp. 417–426.
56. Wang, P., Wu, Q., Shen, C., Dick, A., & Van Den Henge, A. (2017). Explicit knowledge-based reasoning for visual question answering. In *Proceedings of the 26th international joint conference on artificial intelligence*. IJCAI'17. MIT Press Book, USA, pp. 1290–1296.
57. Wang, X., Wang, D., Xu, C., He, X., Cao, Y., & Chua, T. S. (2019). *Explainable reasoning over knowledge graphs for recommendation*. USA: MIT Press Book.
58. Webster, J., & Watson, R. T. (2002). *Analyzing the past to prepare for the future: Writing a literature review*. Management Information System Quarterly, pp. xiii–xxiii.
59. Wohlin, C. (2014). Guidelines for snowballing in systematic literature studies and a replication in software engineering. In *Proceedings of the 18th international conference on evaluation and assessment in software engineering*. EASE '14. New York, NY: ACM, pp. 38:1–38:10.
60. Yan, K., Peng, Y., Sandfort, V., Bagheri, M., Lu, Z., & Summers, R. M. (2019). Holistic and comprehensive annotation of clinically significant findings on diverse ct images: Learning from radiology reports and label ontology. In *The IEEE conference on computer vision and pattern recognition (CVPR)*. IEEE Explorer, Long Beach, CA.

6 The Metadata Management for Semantic Ontology in Big Data

Farhana Kausar, Devi Kannan, Mamatha T. and Aishwarya P.

CONTENTS

6.1 INTRODUCTION

According to National Information Standards Organization (NISO)[1] "Metadata is structured information that describes, explains, locates, or otherwise makes it easier to retrieve, use, or manage an information resource. Metadata is often called data about data or information about information." Metadata, which literally means detailed information about "data about data," has become a widely used but sometimes misunderstood term that is interpreted differently by the various professional communities involved in the information systems and the resources that are designed, created, described, preserved, and used. It is a concept that's been there for as long as people have been organizing information, invisible in many circumstances, and metadata has become more obvious as there is increase in creating and interacting with the digital era.

Metadata construction and management has traditionally been in the realm of cataloguing, categorization, and indexing professionals; but, as more information resources are made available, metadata considerations are now accessible to everyone who uses the Internet. People that are adept at developing and utilizing networked digital content employ metadata. As a result, it is now more important

DOI: 10.1201/9781003310792-6

than ever that everyone who creates or uses digital content, including information professionals, understands the crucial roles that different types of metadata play in these processes. These processes ensure that information is important and that record-keeping systems are accessible, authoritative, interoperable, scalable, and maintainable. Regardless of an information object's physical or intellectual form, metadata can and should transmit three properties: content, context, and structure.

- An information object's content is crucial since it states what the entity contains or is about.
- Context describes who, what, why, where, and how various factors have impacted an entity's development and is knowledge that exists independently of an information item.
- Structure can be internal, external, or a combination of the two. It is the formal set of relationships that exist inside or among specific information elements.

Types of Metadata:

- **Administrative metadata**: Administrative metadata contains information that can be used to manage a resource, such as when and how it was generated.
- **Descriptive metadata**: Descriptive metadata, such as title, abstract, and author, provide information about the source.
- **Structural metadata**: Information required to capture an item's internal structure so that it can be shown to the user in a logical format (for example, a book must be supplied in its page order). This form of metadata is required since an item may contain many (sometimes thousands) of files.

6.2 LITERATURE SURVEY

Business and information technology (IT) leaders should pay attention to three big data characteristics outlined by Gartner [2] in 2011: Information volume,[3] diversity,[4] and volume.[5] While "volume" refers to increasingly different data kinds, such as tabulated (database), hierarchical (hierarchical data), documents, e-mail, metering data, video, photographs, music, stock ticker data, and financial transactions, it must be managed; this is referred to as "rising variation." The rate at which data is generated and the rate at which it must be processed in order to satisfy stakeholders' needs are both referred to as "velocity." The International Business Machines Cooperation (IBM) adds "veracity" to the volume, variety, and velocity aspects to measure data trustworthiness because datasets come from a variety of places. Modern product information data sets are connected and complicated, so new methods of gaining value from big data will be needed. Handling large data encompasses analysis, capture, querying, and visualization when building a technique or programme for big data administration. Sharing, storing, visualizing, and updating [6] are all options, with appropriate modeling and simulation algorithms, in particular. Different types of data are represented, as well as individually adjustable data connection technologies. For such an approach, dynamic datasets are useful.

Semantic technology can facilitate interpretation by identifying the right ideas and context when producing and conveying complex information and interactions between concepts. Using semantic technology, humans and machines can communicate the intent and meaning of data (such as symbols, phrases, etc.) and complex concepts.[7] Semantic technologies can be created using metadata, which contains additional information about other data and that speeds up the process of finding information and papers. Additional metadata from various data sources can also be attached to metadata. This, on the other hand, needs well-defined norms for the formal encoding of metadata. The transmission of metadata between information systems is subject to certain constraints. Previous work in these domains has shown that semantic technologies can be used in the field of information systems, notably for product data. For the purpose of enabling the modeling of eClass catalogues as an ontology, Hepp (2005) introduced a special ontology vocabulary called eClassOWL.[8] Making an ontology that represented data from an eClass catalogue was the aim of this research. Brunner (2007) offered an overview of how semantic web technologies could enhance product data management information systems.[9] A product information system based on semantic technologies was also demonstrated by Brunner et al. But they came to the conclusion that the available semantic web technologies at the time did not permit the creation of an effective and scalable product information system. As a result, custom upgrades were necessary.

The core business entities that a corporation routinely employs across numerous business processes and systems are referred to as master data [9]. The most significant type of master data is product information, and product information management (PIM) is increasingly vital for contemporary businesses since it offers a rich business context for numerous applications. Existing PIM systems are not strong enough to fully capture and apply the semantics of master data and, thus, are less adaptable and scalable for on-demand business.

A basic catalogue layout, BMEcat is a companion ontology to the previously described Good Relations ontology. J S Brunner and his colleagues presented a comprehensive master data management strategy [9]. The use of the same data structures across the product life cycle is the focus of this method. The preliminary results of a research strategy are presented in the form of a semantic master data management reference architecture.

A traditional product information management system, also known as product master data management (P-MDM),[10] needs to be complemented by other MDM domains in the context of a larger product life cycle management environment. These MDM domains might include ones for events, customers, finances, suppliers, and other things. A genuine cross-enterprise semantic integration capability is needed to address such a wide range of requirements. The currently employed commercially available technologies are unable to meet this capability. In order to create a reference architecture and a multi-domain ontology, this study of multi-domain ontology recommends a research approach that would draw on research and development initiatives, particularly in ontology engineering, in both academic and industrial domains. According to Shvaiko [11], it is appropriate to ask various issues about ontology matching after years of study, such as if the area is still developing. Is this development important enough to warrant continuing the study? If so, which directions appear to be the most promising? We examine the current state of ontology matching technology and examine the findings of recent ontology matching evaluations in order to provide answers to these issues. These findings demonstrate meaningful advancement in the discipline, albeit at a slower rate. We hypothesize that large advancements in ontology matching can only be made by tackling fundamental difficulties. We address these issues and offer suggestions for solutions in an effort to focus research on the most promising areas and advance the discipline.

The authors [11] insists that the semantic web [SVD] is widely acknowledged as an excellent infrastructure for increasing knowledge exposure on the web. Ontology, which is used to formally reflect human conceptualizations, is the foundation of the semantic web. According to author,[12] the languages like RDF, RDFS, and OWL are largely used to facilitate ontology engineering in the semantic web. This chapter explores ontology requirements in relation to the web, contrasts the aforementioned three languages with current knowledge representation formalisms, and examines ontology management and application tools. Three real-world applications are used to show the benefits of employing ontologies in knowledge-base-style and database-style systems. The authors point out that there is currently no guiding principle for managing metadata in big data ecosystems.[13] Large enterprises find it challenging to share data, integrate data, write code for data preparation and analysis, and ensure that analytical code makes assumptions that are consistent with the data it utilizes. This use-case chapter outlines the difficulties and a current initiative to deal with them. We give an example of a real application, talk about the criteria for “big metadata” derived from that example and other analytical applications used by the US government, and briefly talk about an effort to modify an open-source metadata manager to accommodate big data ecosystems.

Due to variables including increasing product complexity, globalization, the emergence of virtual firms, and more client focus, a more comprehensive and rigorous approach [14] to managing knowledge is necessary. For corporate organizations, human resource management, and organizational culture, knowledge management is a major issue. IT is a vital ally in managing knowledge. The authors suggest both a method for developing ontology-based knowledge management (KM) systems and a collection of tools for doing so. The contrast between knowledge process and knowledge metaprocess is the foundation of their methodology. They use CHAR, a knowledge management system for corporate history research, to demonstrate their processes. Analyzing KM procedures and dynamically improving the KM system are difficult tasks. The framework created by the authors is intended to aid in leveraging these changing systems by offering a clear understanding of the issue.

This chapter presents a unified framework of big data and unveils ten significant features (10 Bigs) of big data, and then analyses their non-linear interrelationships.[15] The three layers of the framework are the fundamental, technological, and socioeconomic levels. Big data, at its most basic level, includes four major core properties. Three major big data technology qualities make up the technological level. Three significant socioeconomic aspects of big data are present at the socioeconomic level. The chapter takes a service-oriented approach to examining each level of the suggested architecture. The research and development of big data, big data analytics, data science, business intelligence, semantic ontology, and business analytics may be aided by the methods suggested in this chapter.

High levels of heterogeneity [16] and complexity define the Internet. New and exciting data mining use cases may be made possible by the emerging metadata. The use of servers helps to display the complex data in a straightforward manner. Currently, the Transport Layer Security (TLS) on the Internet is the default standard for encrypted communication. Due to ongoing development and the need for backward compatibility, it has developed into a complicated ecosystem.[17, 18] As a result, the protocol already offers a variety of metadata about client and server capabilities that are shared during the first TLS handshake and that may be used to describe a server. TLS fingerprinting has been proposed as a technique to offer real-time enforcement by disclosing the identity of the starting process after analyzing the client's initial TLS packet.[19] Traditional TLS fingerprinting extracts the metadata from the TLS client greeting and generates a fingerprint string using a predetermined protocol. These techniques work with all TLS protocol versions, including TLS 1.3, where the required data characteristics are still provided without encryption. TLS fingerprinting converts a fingerprint string into a mapping to a process using a dictionary of pre-existing fingerprint-to-process mappings. Unfortunately, TLS fingerprint strings, which commonly map to tens or hundreds of different processes, frequently indicate a TLS library rather than a specific process.

6.3 WHY IS METADATA MANAGEMENT NEEDED?

Metadata management is critical because it allows you to better analyze, aggregate, categorize, and sort data using metadata. Metadata is also to blame for a lot of data quality difficulties. Metadata means that in the future, we will be able to locate, use, keep, and use data, which makes it much easier to find relevant data. Because most searches (like a Google search) are language-based, media like music, photos, and video are limited unless text metadata is included. Metadata also helps with text document discovery by characterizing the document's content. Before they can use the data, researchers must understand how it is formatted, the language used, how it was gathered, and how to comprehend it. When it comes to data reuse, researchers frequently wish to incorporate data from a previous study into their own. The data must still be located and used, but with greater assurance and comprehension. Data reuse typically necessitates meticulous documentation and storage. It's also an excellent way to organize electronic resources, which is critical given the increasing number of web-based resources. Traditional resource links were organized as lists and published as static webpages with hardcoded names and uniform resource locators (URLs) in hypertext markup language (HTML). Using metadata to assist in the creation of these pages, on the other hand, is a more efficient way. Software applications can be used to retrieve and convert data for use on the web. Metadata is used to support resource interoperability and integration. When used to describe resources, metadata can be interpreted by both humans and machines. This makes it possible for systems with various operating systems, data formats, and interfaces to exchange

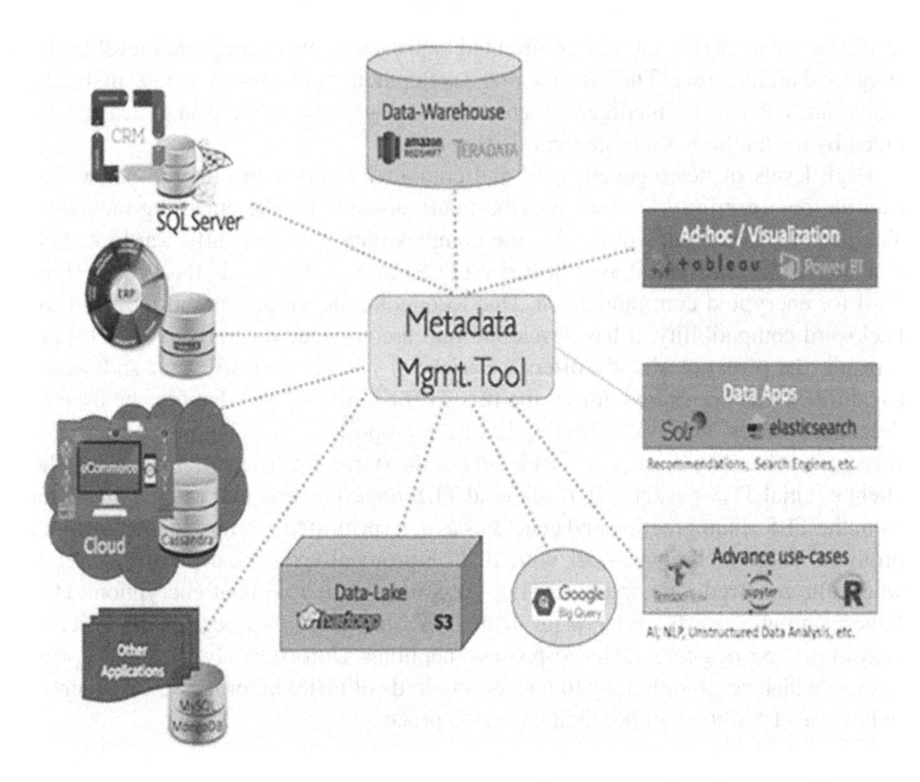

FIGURE 6.1 Representation of a metadata tool at different levels of architecture.[20]

data with each other while maintaining the highest level of interoperability. As a result, it is now much simpler to find resources across the network.

Figure 6.1 shows the different aspects of infrastructure [20] which are connected with the metadata management tools. This tool helps to supplement solutions to different sources like data warehouses, backend servers, and cloud and other applications for future use. The proper storage of metadata helps in retrieving and storing and searching becomes easier to locate and understand. Figure 6.2 shows the lineage of data across different platforms.

6.4 CATEGORIES OF DOCUMENT METADATA

The key features of document metadata are assisting, organizing, and quickly finding documents.

- **Properties:** Based on the document kinds or templates,[19] the document maker will be able to input precise information. Metadata fields are used to provide additional information for the purposes of organizing and locating material. Fields include things like names, dates, numbers, and money. They can be customized to meet your needs.

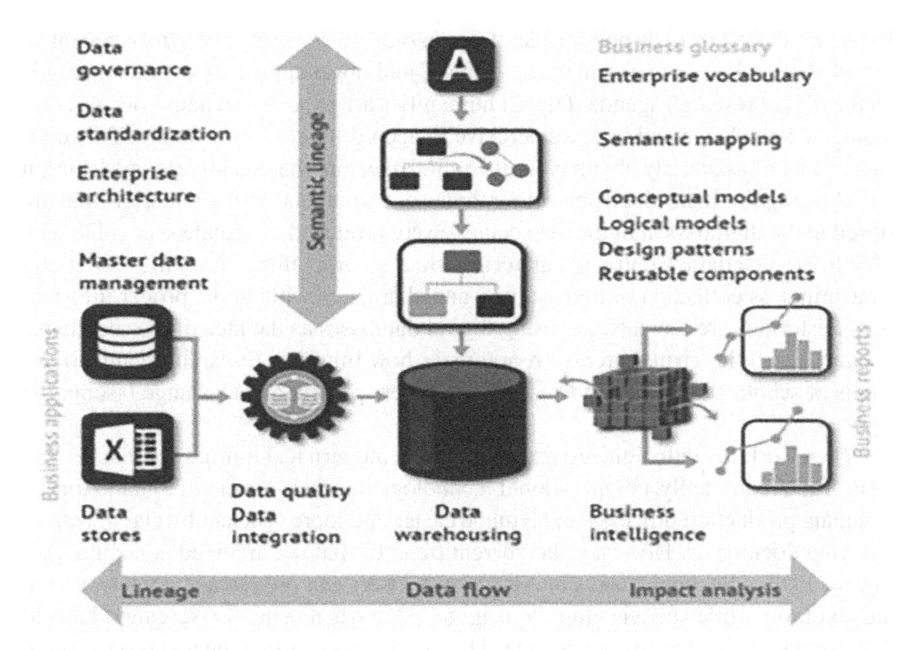

FIGURE 6.2 Lineage of data across different platforms from data storage to providing intelligence to business.[19]

- **Version related information:** Version-related data, such as the version number, revision, creation date, modified date, and so on, aid in version number, revision, creation date, and so on. For auditing purposes, this parameter is critical.
- **Comments:** The collaborative process is aided by commenting on the document. Top systems allow users to make comments to each version. Workers can tag each other for cooperation invitations and notice, and a remark can be a text with attachments. Searching for papers based on comments may also assist you in finding them when you need them.
- **Tags:** They are used to sort documents into categories. They are frequently defined at the document level, which means they can't be changed between versions.

6.5 THE IMPORTANCE OF CREATING METADATA FOR BIG DATA

Digital humanities [3] have evolved as a new paradigm in the last decade; they bring together scholars interested in using computer approaches on their study materials. The virtually exponential cxpansion of either born-digital or digitized materials currently available for scholars has aided this evolution. Furthermore, computational research tools are far more readily available today than they were a decade ago.

Increasingly common terms like "big data," "data mining," and "text mining" graphically demonstrate the vast amount of digital data that is accessible for research.

However, there are high possibilities for exploratory research, the advancement of knowledge in the social and natural sciences, and novel approaches to data analysis on the digital research agenda. Digital humanities are expected to alter how we communicate knowledge and how we perceive it, according to some scholars. This division of labor has already changed in the digital age, and there is no reason to think it will do so again. Instead, billions of terabytes of textual and visual materials that are stored in the digital form in the data centers were produced and made accessible with little to no metadata. Contrarily, materials are a separate thing. They are essentially a meaningless collection of files, values, and characters without the proper information. Historians are increasingly using digital databases as the idea of the searchable document and the virtual archive reorganizes how libraries, research organizations, teams of scholars, and even individual researchers present and exchange fascinating sources.

When working with digitized textual corpora, modern text mining approaches can be useful. Additionally, computational technologies such as (semi-) automatic for the metadata production process can be made easier and more efficient by classifying or indexing documents. However, the current trend of making archived content accessible as PDF collections makes matters worse. Users can browse and copy text from the text layer while still viewing the original page when using the so-called layered PDF format, which has this advantage. The disadvantage is that the text layer is often a character-based reconstruction of the page rather than a raw text output paginated in accordance with the original design (based mostly on corrected optical character recognition [OCR] results). Since the first sentence ends in a hyphen, hyphenated words on two lines are regarded as two separate sentences. Even if the research interface contains more complex search capabilities, such as regular expressions, we can picture the kinds of limitations that come with this method to document discovery. This is the case due to the fact that most search engines are built to match patterns, irregularly divided words, for instance, lack a recognizable pattern.

6.6 USING ONTOLOGY IN SEMANTIC WEB AND BIG DATA MANAGEMENT

The World Wide Web has been a huge success as an information technology. It has changed how information [21] is produced, stored, and shared in a variety of industries, including retail, family photo albums, and high-level academic research, in only 10 years. Although it has not yet had anything close to the Web's exponential acceptance, the semantic web is being hailed as revolutionary by its creators. It aims to go above the web's existing limitation, which is that indexing is mainly restricted to certain character strings. As a result, even if a person searches for information on "turkey" for the bird, they will find a plethora of meaningless sites about "Turkey" that are about the country where the speaker is capable of comprehending such material. Most experts believe that semantic web apps will need to use some type of shared, organized, machine-readable conceptual architecture in order to be viable. A formal ontology is being created as a result of a collaboration between the semantic web research community and an earlier school with origins in traditional artificial intelligence (AI) research (also known as "knowledge representation"). The most fundamental ideas or "categories" necessary to comprehend data from any knowledge domain are described in a formal ontology, which is a machine-readable description.

The big data management [4] is no longer just a problem for big businesses; it's also a problem for small and medium-sized businesses. Regardless of their size, businesses today must manage more complex company data and procedures that are almost totally electronic in nature. To minimize and analyze the complexity of company data and processes, enterprise information systems require functionalities based on specialized technologies. The purpose of this research is to investigate how semantic technologies, particularly ontologies, might be used to improve current information systems. For this aim, three examples of potentially improved information systems are given, along with methods based on ontologies and semantic technologies. Data integration, data quality, and business process integration are a few of the use cases that are covered. The usage of ontologies, [22] which foster interoperability and common understanding between participants, is a crucial component in overcoming semantic heterogeneity and encouraging semantic interoperability between various online applications and services. Recently, ontologies have been popular in a variety of domains, such as natural language processing, knowledge engineering, electronic commerce, and knowledge management. Ontologies offer a shared concept of a subject that can be conveyed between people as well as a common knowledge of a variety of frequently used application systems. They were designed to make knowledge interchange and reuse easier in research communities for AI.

Taxonomies and ontologies are related, but ontologies include additional meaningful links between concepts and attributes in addition to precise rules for describing concepts and relationships. The skeleton of a knowledge base is an ontology, which is a set of concepts used to characterize a topic and is organized hierarchically. According to this theory, it is possible to create several knowledge bases with the same taxonomy or skeleton by using a single ontology.

The ontology community makes a distinction between ontologies that are mainly taxonomies and those that provide a more thorough explanation of the domain and impose additional restrictions on its semantics. They are referred to as lightweight and heavyweight ontologies, respectively, in the community. Lightweight ontologies are made up of ideas, concept taxonomies, relationships between concepts, and attributes that identify concepts as opposed to heavyweight ontologies, which include axioms and boundaries. These aid in understanding the language of the ontology. A variety of knowledge modeling techniques can be used to develop and implement ontologies in both heavyweight and lightweight languages.[23] Following are some different types of ontologies:

- If they are expressed in natural English, they are exceedingly informal. An excessively casual ontology, according to this, would be unacceptable. It's called an ontology since it cannot be read by machines.
- When stated in a machine-readable form of natural language; when represented in an artificial and formally defined language (such as RDF graphs); and when represented in an artificial and formally defined language (e.g., RDF graphs).

- If they contain explicitly specified notions in formal language, such as semantics, theorems, and evidence of soundness and other qualities, they are rigorously formal (web ontology language [OWL]).

Prior to OWL, a lot of work was made towards creating a potent modeling language for ontologies.[24] RDF and RDF/S based on XML were the starting point for this development, which eventually transitioned to the ontology inference layer (OIL). As stated by Taye [12] there are a few key requirements for high-quality support that should be taken into consideration while creating languages for encoding ontologies. A text format that is simple to read, expressive power, compatibility, sharing and versioning, internationalization, formal conceptualizations of domain models, precisely defined syntax and semantics, effective reasoning support, adequate expressive power, and user-friendly expression are some of these qualities. One of a language's most crucial elements is its syntax, which needs to be precisely defined. Syntax is the most crucial criterion for computer data processing. One of the most important features of any language is its syntax, which should be well-defined; it is also the most important need for computer data processing. The following are the big data applications where semantic ontology is applied with the help of metadata management.

The semantic web:[25] The semantic web relies heavily on ontology to facilitate information flow across remote domains because the semantic web is seen as an extension of the current web since it represents data in a machine-processable way. The discovery of Semantic Web Services[25] in the world of e-business, ontology is important since it helps find the best match for a requester looking for products or other objects. Additionally, it facilitates e-commerce businesses.

Artificial intelligence:[25] Ontology was created in the AI research community with the purpose of supporting knowledge sharing, reuse, and processing across a domain amongst programmes, services, agents, or organizations.

Big data still needs a well-thought-out, comprehensive data management strategy. Rethinking information management in the context of big data technology should be the focus of future research. Examples include documentation, reconfiguration, data quality assurance, and verification. Many critical jobs are difficult to support in the existing environment technology for big data. Representing data with appropriate schemata was a landmark milestone for the database community. Unfortunately, because big data deals with continually changing heterogeneous data, determining a data structure prior to processing is difficult, if not impossible. There was a lot of interest in schema-less data integration and access solutions.[26]

However, because schema-based representation techniques are difficult to apply to big data. The representation of metadata within data-intensive systems is receiving increasing attention. To manage a significant number of heterogeneous and distributed data, metadata representing diverse data documentation, provenance, and trust are all facets of semantics and data quality, correctness, and other qualities, that must be defined and updated on a regular basis.[26] The Institute of Electrical and Electronics Engineers (IEEE) recently started a project targeted at promoting big data standardization technologies. Data processes and flows can be modelled, which requires describing the complete pipeline and making data representations shareable and verifiable. This is another way to apply the concepts of data semantics to large data. To process a data stream, for instance, the right analytics and data preparation modules must be selected. Last but not least, we want to underline how significant issues with data quality are. For big data analytics to produce high-quality results, each step of the pipeline has its own set of quality responsibilities that must be taken into account.

6.7 MAJOR CHALLENGES

The following are the major challenges for semantic ontology in big data management.

Developing a query language capable of handling complicated, heterogeneous and incomplete data is a challenge. Furthermore, user preferences tend to be critical to provide in a query, such as service quality, data quality, output data format, and desired display method. Another point of interest is declarative query support, as well as query optimization. When it comes to defining sophisticated analytical tasks, SQL is generally seen as inadequately powerful. As a result, data scientists prefer domain-specific languages (DSLs) that mix programming languages with domain-specific languages like Python, Pig,[27] special-purpose languages like R,[28] or user-defined functions to execute specialized tasks. Parallelizing and optimizing sophisticated dataflow systems is tough with imperative languages like Python or Java, and black-box UDFs in particular.

Creating models that include regulatory knowledge to automate compliance is another difficulty. It is impossible to tackle regulatory quirks project by project. Instead, all parties involved in big data analytics should have access to each big data project's certified compliance (e.g., in the form of a privacy impact analysis). Furthermore, data processing creates legal issues, which could lead to an undesirable lawsuit. How do your account for intellectual property and affect the economic exploitation of analytics in multiparty scenarios?[28] How can evidence be presented that data processing is ethical[29] other rules and guidelines? These are some of the difficulties for which mature and effective answers are still needed.

With a focus on providing distributed descriptions (commonly referred to as annotations) to web resources or applications, metadata research has arisen as a topic that cuts across many different disciplines. In a variety of application fields, including search and location, personalization, federation of repositories, and automated information distribution, these successive descriptions are meant to serve as the basis for more complex services. Ontology-based data has a real technological foundation thanks to the semantic web. For instance, complex and extensive metadata schemas are needed for big biological databases to enable more accurate and knowledgeable search tools, whereas web-based social networking calls for metadata describing individuals and their relationships. The languages and idioms used to offer meta-descriptions are very diverse, ranging from simple structured text in metadata schemas to formal annotations using ontologies. The technology used to store, share, and use meta-descriptions are similarly very diverse and quickly developing. The development of specialized knowledge and abilities in this area is required since the expansion of schemas and standards connected to metadata has also produced a complex and dynamic technical environment.

6.8 CONCLUSION

An integrated knowledge base that can be transferred from individuals to application systems is what an ontology aims to provide. Because they try to capture domain knowledge and have as their goal the unambiguous general expression of semantics, ontologies are essential for facilitating interoperability between organizations and the semantic web, and they are the basis for domain agreement. As a result, in various circles, ontologies have become a hot topic. The concept, structure, major operations, and uses of ontology are all discussed because ontology is such a crucial aspect of this subject. For ubiquitous and portable machine understanding, the semantic web strongly relies on formal ontologies to structure the underlying data. The distribution of ontologies is therefore crucial to the success of the semantic web, necessitating quick and easy ontology engineering in order to avoid a delay in knowledge acquisition. Ontology learning greatly facilitates the ontology engineer's ability to create ontologies. We provide a vision of ontology learning that integrates a range of complimentary disciplines that feed on various types of unstructured, semi-structured, and fully structured data in order to support a semi-automatic, collaborative ontology engineering process. The ontology learning framework gathers a wealth of coordinated data for the ontology engineer by importing, extracting, pruning, recursively incrementing, and assessing ontologies. As a result, the processing of unstructured and semi-structured data, as well as the choice of data structuring levels, are closely related to metadata. Although the metadata used to describe web URLs is typically semi-structured, it does adhere to some standards and consistent models that ensure operational interoperability in a diverse context.

The design process has gotten more challenging due to the complexity of big data applications and the absence of standards for modeling their constituent parts, computations, and processes. Developing data-intensive apps is a risky and resource-intensive endeavor. We argued in this chapter that there is no such thing as innovation. Lack of sound modeling practices can be compensated by algorithms. Indeed, we believe that the major challenges that we are confronted with are from emergence of different technologies even more than inventing new technologies, big data research necessitates developing novel data management strategies through analytics capable of delivering non-functional characteristics, such as data quality. Whether it is data integration, model compliance, or regulatory compliance, we've got you covered. Such issues can be addressed through data semantics research. Future research should be conducted in accordance with the FAIR principles for putting in place design procedures that result in findable data that is accessible, interoperable, and reusable.

REFERENCES

1. National Information Standard Organization. https://niso.org/
2. Gartner's Big Data definition, gartner.com.
3. Kimmo, Elo. (2020, Dec.). *A methodological note on the importance of good metadata in the age of digital history, digital histories: Emergent approaches within the new digital history publisher*. Helsinki University Press, Mountain View, California, USA.

4. Eine, B., Jurisch, M., & Quint, W. (2017, May). *Ontology-based big data management.* MDPI © 2017 by the authors. Licensee MDPI, Basel, Switzerland.
5. Dumbill, E. What is big data? Retrieved May 10, 2017 www.oreilly.com/ideas/what-is-big-data. The 2012 Strata Conference, being held Feb. 28-March 1 in Santa Clara, California
6. Labrinidis, A., & Jagadish, H. V. (2012). Challenges and opportunities with big data. *Proceedings VLDB Endow*, 5, 2032–2033. [CrossRef]
7. Maedche, A., & Staab, S. (2001). *Learning ontologies for the semantic web.* Hongkong, China: Semantic Web Workshop.
8. Hepp, M. (2005, November 6–10). eClassOWL: A fully-fledged products and services ontology in OWL. In *Proceedings of the 4th international semantic web conference (ISWC).* Galway, Ireland. E-business + web science research group.
9. Brunner, J. S., Ma, L., Wang, C., Zhang, L., Wolfson, D. C., Pan, Y., & Srinivas, K. (2007, May 8–12). Explorations in the use of semantic web technologies for product information management. In *Proceedings of the 16th international conference on world wide web.* ACM, Banff, AB, Canada, pp. 747–756.
10. Fitzpatrick, D., Coallier, F., & Ratté, S. (2012, July 9–11). A holistic approach for the architecture and design of an ontology-based data integration capability in product master data management. In *Proceedings of the product lifecycle management: Towards knowledge-rich enterprises—IFIP WG 5.1 international conference*, vol. 388. Springer, Montreal, QC, Canada, pp. 559–568. Product Lifecycle Management. Towards Knowledge-Rich Enterprises: IFIP WG 5.1 International Conference, PLM 2012, Montreal, QC, Canada, July 9–11, 2012.
11. Shvaiko, P., & Euzenat, J. (2011). Ontology matching: State of the art and future challenges. *Transactions on Knowledge and Data Engineering*, 25, 158–176. [CrossRef]
12. Taye, M. (2009). Ontology alignment mechanisms for improving web-based searching. Ph.D. Thesis, De Montfort University, UK.
13. Smith, K., Seligman, L., Rosenthal, A., Kurcz, C., Greer, M., Macheret, C., Sexton, M., & Eckstein, A. (2014). Big metadata: The need for principled metadata management in big data ecosystems. In *Proceedings of workshop on data analytics in the cloud, series DanaC'14.* ACM, New York, pp. 13:1–13:4. https://doi.org/10.1145/2627770.2627776
14. Staab, S., Studer, R., Schnurr, H. P., & Sure, Y. (2001, January). Knowledge processes and ontologies. *IEEE Intelligent Systems*, 16(1), 26–34. https://doi.org/10.1109/5254.912382
15. Sun, Z., Strang, K., & Li, R. (2018, October). Big data with ten big characteristics. In *ICBDR '18: Proceedings of the 2nd international conference on big data research.* Association for Computing Machinery, New York, pp. 56–61. http://dx.doi.org/10.1145/12345.67890
16. Anderson, B., & McGrew, D. Accurate TLS fingerprinting using destination context and knowledge bases. arXiv:2009.01939v1
17. Kotzias, P., Razaghpanah, A., Amann, J., Paterson, K. G., Vallina-Rodriguez, N., & Caballero, J. (2018). Coming of age: A longitudinal study of TLS deployment. In *Proceedings ACM international measurement conference (IMC).* ACM, IMC '18, October 31-November 2, 2018, Boston, MA, USA.
18. Sosnowski, M., Zirngibl, J., Sattler, P., Carle, G., Grohnfeldt, C., Russo, M., & Sgandurra, D. Active TLS stack fingerprinting: Characterizing TLS server deployments at scale. arXiv.2206.13230v1. U Parkhotel University of Twente, Enschede, The Netherlands.
19. www.talend.com/resources/metadata-management-101/
20. Varshney, S. (2018, March). A step by step guide to metadata management. *Big Data, Metadata Management.*
21. Mohammad Mustafa Taye. (2010, June). Understanding semantic web and ontologies: Theory and applications. *Journal of Computing*, 2(6), ISSN 2151–9617.

22. Mihoubi, H., Simonet, A., & Simonet, M. (2000). An ontology driven approach to ontology translation. In *Proceedings of DEXA*. Springer, pp. 573–582, 11th International Conference, DEXA 2000 London, UK, September 4–8, 2000 Proceedings.
23. Ehrig, M., & Euzenat, J. (2004). *State of the art on ontology alignment*. University of Karlsruhe: Knowledge Web Deliverable D2.2.3.
24. Ding, L., Kolari, P., Ding, Z., & Avancha, S. (2007). *Using ontologies in the semantic web: A survey, integrated series in information science*. Springer, Boston, MA.
25. Berners-Lee, T., Hendler, J., & Lassila, O. (2001, May). The semantic web. *Scientific American*, 34–43
26. Franklin, M. J., Halevy, A. Y., & Maier, D. (2005). From databases to dataspaces: A new abstraction for information management. *SIGMOD Record,* 34(4), 27–33. https://doi.org/10.1145/1107499.1107502
27. Olston, C., Reed, B., Srivastava, U., Kumar, R., & Tomkins, A. (2008, June 10–12). Pig Latin: A not-so-foreign language for data processing. In *Proceedings of the ACM SIGMOD international conference on management of data*. Vancouver, BC, Canada: SIGMOD, pp 1099–1110. https://doi.org/10.1145/1376616.1376726
28. Venkataraman, S., Yang, Z., Liu, D., Liang, E., Falaki, H., Meng, X., Xin, R., Ghodsi, A., Franklin, M. J., Stoica, I., & Zaharia, M. (2016, June 26–July 01). Sparkr: Scaling R programs with spark. In *Proceedings of the 2016 international conference on management of data, SIGMOD conference 2016*. San Francisco, CA, pp 1099–1104. https://doi.org/10.1145/2882903.2903740
29. Martin, K. E. (2015). Ethical issues in the big data industry. *MIS Quarterly Executive*, 14(2).

7 Role of Knowledge Data Science during Covid-19

Mokshitha Kanchamreddy and
Karthika Natarajan

CONTENTS

7.1 INTRODUCTION

Data science is "an intersectional analysis that integrates scientific approaches, strategies, processes, and techniques to obtain knowledge Includes information from numerous sources construction and incomplete datascience". For the past few years, we have been facing an unpredictable situation with the coronavirus outbreak. In this type of situation, the data science community tries to find simple and easy ways to predict and analyse the outcome of an outbreak. Moreover, specialists from other disciplines and federal agencies understand the importance of data analytics by making the virus' DNA and other datasets publicly available in the hopes of developing an artificial intelligence (AI)-guided cure.

Data science is concerned with the collection, analysis, and decision-making of data. Finding patterns in data, analysing it, and making future predictions are all

DOI: 10.1201/9781003310792-7

part of data science. Companies can use data science. The pandemic's rapid spread, including its ever-changing patterns and varying symptoms, makes it extremely difficult to contain. Furthermore, the pandemic has impacted health systems and medical resource availability in several nations around the world, leading to the high fatality rate.[1]

Periodic check-up of people is planned, and suspected Covid-19 instances will be quickly tracked with the help of remote detection technologies. Additionally, the utilisation of such systems will produce a significant amount of data, which will present a number of opportunities for utilising big data analytics approaches to raise the calibre of healthcare services. There are a huge number of open-source software tools that are made to operate in a distributed and cloud computing environment and can aid in the creation of big data-based solutions, such as the big data components of the Apache project.

Analysing the widespread accessibility of big data from human mobility, contact tracing, medical imaging, virology, drug screening, bioinformatics, electronic health records, and scientific literature, combined with ever-increasing computing power, has made the use of data science methodologies in medicine and public health possible. Data science has become more important than ever in understanding and containing the ongoing Covid-19 pandemic because of these advancements and the intense zeal of academics.

As of 19 May 2021, Covid-19, which is brought on by the severe acute respiratory syndrome coronavirus 2 (SARS-CoV-2), has infected nearly 3.4 million people worldwide. The Covid-19 pandemic emphasises the urgent need for timely and reliable data sources that are both individualised and population-wide to generate data-driven insights into disease surveillance and control because of its tremendous influence on global health and economics. This is in contrast to responses to epidemics like SARS, Ebola, HIV, and MERS.[2]

For mathematicians, physicists, and engineers, they may contribute to the knowledge of diseases from data-driven and computational perspectives by using the Covid-19 pandemic as a platform and a rich data source. When previous epidemics occurred, some of these data were unavailable, while others were accessible but hadn't yet reached their full potential. It became clear that traditional public health measures had failed to contain the virus within a fleeting period of time after it was first discovered. Looking back, the public health systems had several obvious flaws. Even if it was challenging, the last year saw a notable increase in interdisciplinary, data-driven research on new infectious illnesses. Therefore, it is crucial to assess the achievements to date and to draw out a roadmap for an emerging area that uses data science and cutting-edge computer models to combat future infectious diseases.

The ongoing battle against Covid-19 and upcoming diseases will require a decisive weapon: data science. The selecting process is outlined as follows. We started by searching the Web of Science by Clarivate Analytics for any publications that were pertinent from 1 January 2020 to 31 May 2021 using the keywords ('COVID-19' *OR '2019-nCov') *AND ('data science' *OR 'artificial intelligence'). Second, we conducted a DBLP (a computer sciences bibliographic database) search using the same keyword combinations to locate other conference papers. Third, we used the journals' impact factors and citation counts to rank the retrieved papers. Clinical studies, integrated analysis of pharmacy and insurance claims, and the creation of biomarkers are some novel and inventive methods for analysing healthcare big data.[1–4]

7.2 TECH FOR ANALYSING DATA

Emergency rooms and medical supplies like ventilators saw an increase in demand throughout the Covid-19 pandemic. As an outcome, various research has worked to create models and methods for monitoring that might help in making a variety of medical decisions to lessen potential risks. These solutions include the following. A patient monitoring platform created by the authors enables daily electronic symptom monitoring, text message guidance and reminders, and phone-based care. The amount of people who are trying to procure data has exploded the production of electronic data, which can be used in a very practical and effective way. Let us consider the rapid increase of creating data in the past few years: A close 1.8 trillion terabytes of data were generated in 2011, according to estimates. We generated more than 2.8 trillion gigabytes of data in only 1 year, in 2012; by 2020, this amount will have increased to 40 trillion gigabytes. The nexus of the three important fields described before is data science. Data must be accessible through computer programming if we are to learn anything from it. Is it preferable to develop a PDF of data or an app where users can type in numbers and obtain a quick prognosis. If we are constructing a model to predict heart attacks in patients? (*learn more* [5-6])

An electronic medical record (EMR), like an electronic health record (EHR), keeps track of the usual clinical and medical information obtained from patients. Medical practise management software (MPM), personal health records (PHR), electronic health records (EHRs), and many other healthcare data elements offer the potential to lower healthcare expenditures while also increasing the quality and efficiency of services. Healthcare big data consists of payer-provider data (such as EMRs, pharmacy prescriptions, and insurance records) as well as genomics-driven studies (such as genotyping, gene expression data), as well as additional data gathered from the Internet of Things (IoT) and the smart web. EHR adoption was sluggish at the start of the twenty-first century, but it significantly increased after 2009.

The administration and use of such healthcare in monitoring tools. Information technology has become more and more important for the management and use of such healthcare data. A real-time biomedical and health monitoring system has sped up the development and use of wellness monitoring tools and related software that can send alarms and communicate patient health information with the appropriate healthcare practitioners. These gadgets are producing enormous amounts of data, which can be evaluated to deliver clinical or medical treatment in real time. Big data from the healthcare industry holds promise for enhancing health outcomes and reducing expenses.[6] Figure 7.1 shows the analysis of the data.

Big data is the enormous quantities of various types of data produced quickly. Instead of improving customer consumption, data obtained from multiple consumer services and resources are mostly utilised; scientific research and healthcare big

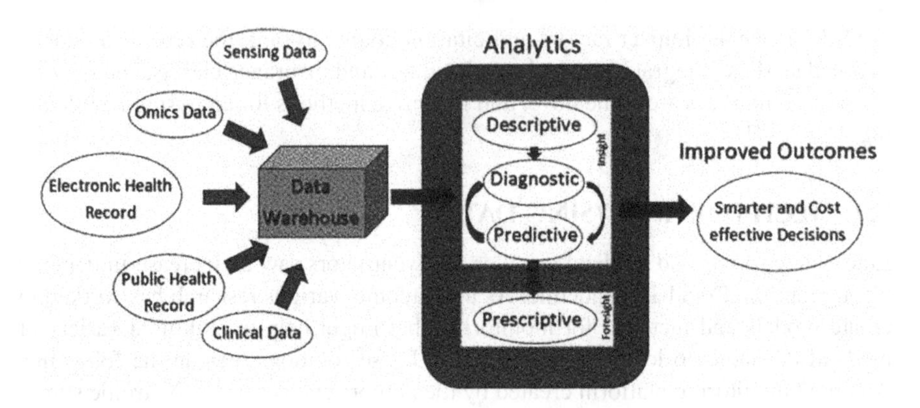

FIGURE 7.1 Picture describing outcomes of data analysis.

data also supports this. Another difficulty in large data analysis is heterogeneity of the data. Using standard technologies, a solution is cloud computing, which offers dependable services and virtualised storage technologies. It delivers great dependability, scalability, and autonomy. These platforms can operate as a computer to analyse and interpret the data, a receiver of data from ubiquitous sensors, and a source of user-friendly web-based visualisation. IoT could do with utilising digital resources, edge devices etc. Sophisticated strategies are required to implement machine learning (ML) and AI approaches for big data analysis on computing platforms. A programming language could be used to develop such algorithms or software, such as Python, R, or another language, that is appropriate for working with huge data. To handle the large data from biomedical research, a solid understanding of biology and IT is necessary. Bioinformaticians typically fulfil this description using more popular technologies for handling. In the next section, we give a brief overview of various platforms.

7.3 EXTRACTING INFORMATION

One of the most important factors in minimising the negative consequences of any epidemic or pandemic is the effective use of information technologies. Current management practises fall short of what is needed to slow the virus's rapid spread. We suggest an infection control system that depends on seamless and prompt information exchange between states and organisations to guarantee efficient resource allocation. This system will make use of AI, epidemic modelling, blockchain technology, and mobile technology. We employed the concept of Multiplatform Interoperable Scalable Architecture to allow for the integration of several platforms and provide a solution for scalability and compatibility problems.

The ability of the healthcare business to process information is being improved by new ML or AI-based tactics. For instance, the field of machine learning known as natural language processing (NLP) is quickly advancing and has the ability to recognise important grammatical patterns in free text, assist with speech recognition, and decipher the meaning of narratives. When dictating clinical notes or creating fresh papers, such as summary of a medical encounter, NLP techniques can be useful. Many NLP developers may find it difficult to handle the complicated nature and distinctive content of clinical documentation. But with the help of techniques like NLP, we ought to be able to extract pertinent information from healthcare data. The predictive power of AI has been used to benefit big data in the realm of medicine. For instance, ML algorithms can automate decision-making using the medical picture diagnosis system.[7]

7.4 ANALYSIS OF IMAGES

There are a lot of methods that are frequently used in healthcare. These methods take large-scale, high-definition medical photographs (patient data). Medical specialists such as radiologists, doctors, and others perform fantastic work in searching for specific problems in these files of medical data. It's crucial to recognise that for many diseases, there are not enough qualified medical personnel. Effective technologies are used in order to make up for this lack of professionals. Especially while the problem is still developing, a medical expert who is preoccupied with detecting a separate condition can miss it. To assist in such circumstances, this method transforms patient diagnosis, therapy, and monitoring by extracting knowledge from vast amounts of clinical imaging data using machine learning and pattern recognition techniques. To perform medical image analysis and find hidden data, numerous software programmes have been developed based on general, identification, recognition, presentation, modelling, and dispersion functions. MRI, fMRI, PET, CT Scan, and EEG are just a few of the different modalities of brain imaging that scientific principles of medicine (SPM) can process and analyse. For instance, the free application Visualization Toolkit gives 3D images from medical tests with advanced processing and analysis. Figure 7.1 provides a summary of many more frequently used tools and their attributes in this field. In order to advance and promote precision medicine projects, such bioinformatics-based big data, analysis may be able to gain more knowledge and value from imaging data for various forms of healthcare and clinical decision support systems. It can be used, for instance, to track novel cancer treatments with specific targets.

7.5 COMMON DATA SCIENCE TECHNIQUES

An ordered and formal relationship between data pieces is referred to as a "data model," which is typically intended to imitate a real-world occurrence. There are many types of data models, including probabilistic and statistical models. The fundamental concept underlying all three topics is that we use data in order to create the best model possible. Both are subsets of a bigger framework called machine learning. We now rely more on statistics than we do on human instincts. Data science is potent because it combines math, computer programming, and subject expertise. When we learn about various modelling techniques, we will understand data science in a better way. Modelling and the health sector can use a variety of data.

7.5.1 Few Techniques Which are Used by Data Science

These days there are lot of techniques that are helping with the current scenario. Let us see some of modelling techniques used.

a) **Linear regression**

An effective method for creating predictions and analysing the interactions between the many aspects of a dataset is linear regression. The optimum outcome of the graph should minimise the sum of all distances between the form and the actual observation. The likelihood of a mistake occurring decreases with decreasing distance between the indicated sites.

Simple linear regression and multiple linear regression are two subcategories of linear regression. Using a single independent variable, the former predicts the dependent variable. The later, however, employs a number of independent factors to forecast the dependent variable and makes use of the best linear relationship.

b) **Non-linear models**

Regression analysis employing observational data that has been modelled by a function is known as non-linear models. When managing non-linear models, data analysts frequently employ a variety of techniques. In data analysis, it is essential to use methods.

c) **Vulnerable vector machines**

Data science modelling methods that categorise data include supported vector machines (SVM). There is a maximum margin established in this limited optimization problem. The limitations used to categorise data, however, affect this variable.

Supported vector machines locate a hyperplane that categorises data points in an N-dimensional space. Data points could be separated by any number of planes, but the trick is to identify the hyperplane that has the greatest distance between the points.

d) **Identification of patterns**

What does pattern recognition mean? You may have heard of this word in relation to AI and machine learning. Technology uses pattern recognition to compare incoming data, which are kept in the database. This data science modelling approach aims to find patterns in the data. Pattern recognition is a subset of machine learning; therefore, it is distinct from that field. It also typically involves two steps. The first step is exploratory, when the algorithms look for patterns without having to meet any specified criteria. In the meantime, the descriptive section's patterns are categorised by the algorithms. Pattern recognition may be used to analyse any kind of data, including text, music, and sentiment. The second stage is that the perceived objects were separated from the background.

e) **Resampling**

Recurring data sampling is a technique used in data science modelling methodologies known as "resampling approaches." The different sample distribution results produced by resampling might be valuable for analysis.

The procedure uses experiential methods to develop the method's characteristic sample distribution.

f) **Bootstrapping**
The performance of a predictive model can be validated using a data science modelling technique called bootstrapping. The technique involves selecting a replacement from the original data using a subset of the data that aren't test cases. Contrarily, cross validation is a different methodology that is used to verify model performance. The training data is divided up into various components to make it work.

g) **Guidelines for improving data science modelling**
For data analysis, the majority of data science modelling techniques are essential. There are, however, a number of workable methods that may be utilised to optimise the data science modelling process in addition to these models for data analysis. Technology, like data visualisation, for instance, can greatly improve the procedure. It is tough to perform any meaningful analysis. Visualisation greatly simplifies the process. The best data analysis can be greatly aided by the use of the appropriate data analytics platform. It is possible to speed up data processing and deliver insights even more quickly with optimised data analytics solutions.[8–10]

7.6 ANALYSING COVID-19 USING DATA SCIENCES TECHNIQUES

Health care analysis can be done through various methods and technologies used in data science.[10]

7.6.1 Six Sigma Technologies

We use a lot of technologies for the utilisation of data so that it is easy to find a way in which both software and hardware can interact with the data. One of the technologies is healthcare information technology (HIT), which gives interaction of computer components to data and information processing. To help achieve goals, which include eliminating waste and raising the standard of care, several hospitals and healthcare institutions have adopted Six Sigma management techniques. Applying Six Sigma techniques in healthcare settings may improve patient care by removing mistakes and variations in operations, streamlining procedures, cutting costs, and more. A defect is something that makes a patient dissatisfied in a medical environment. One example of an issue is a lengthy wait time for a doctor's appointment, but a wrong diagnosis or course of treatment is a significant one.

Understanding the needs and standards of patients can help healthcare organisations provide better patient care. Six Sigma has proven to have a major influence on healthcare administration, logistics, and primary care, resulting in enhanced productivity and effectiveness. Six Sigma can assist healthcare organisations find ways to maximise resources, eliminate waste, and get the results they need to limit costs and increase patient satisfaction. Needs might include speeding up the transfer of patients from the emergency room to a hospital room or improving response times for laboratory procedures. Here are a few healthcare institutions that have effectively included Six Sigma into their attempts to enhance quality:

- Mount Carmel Health System
- Boston Medical Center
- Rapides Regional Medical Center
- The Women and Infants Hospital of Rhode Island

People now have access to more information about healthcare providers than ever before, and the quality of service is being taken into consideration when choosing a provider. In the field of healthcare, the Six Sigma methodology is still in its early stages, despite its widespread use in high-level manufacturing companies like General Electric, Motorola, Texas Instruments, IBM, and others. Six Sigma is becoming the method of choice for many healthcare businesses in the United States and Europe for quality improvement. It is evident from literature assessments that the majority of healthcare industries have used the Six Sigma DMAIC technique both historically and currently.[10]

7.6.2 Internet of Things (IoT)

These resources can connect multiple technologies to offer the chronically ill and elderly a trustworthy, efficient, and knowledgeable healthcare service.

A professional can track and test a variety of characteristics from his or her patients at their individual workplaces or residences. Therefore, a patient may avoid the need for hospitalisation or even travel with early diagnosis and treatment, significantly lowering the cost of healthcare. Wearable health and fitness trackers, detectors, diagnostic equipment for observing physiological parameters, and other devices or medical devices are a few instances of the Internet of Things (IoT) in medical instruments. These IoT gadgets produce a huge amount of information on wellness. We can forecast the condition of the patient and their evolution from a subclinical to a pathological state if we can link merging this info with other easily accessible healthcare data from personal health records (PHRs) or electronic medical records (EMRs). In reality, IoT-generated big data has proved quite helpful in a number of domains by providing better analysis and forecasts. The methods for outbreak are containment. Because of its unique character and the need for cutting-edge hardware and software applications, IoT data processing would necessitate new operating software. We would have to control the real-time data influx from IoT devices and evaluate it minute by minute. By using healthcare systems, employees are attempting to reduce costs and improve the quality of care. These issues can be solved by data techniques like AI and ML that can work without having to be explicitly programmed by the given issue calls for domain expertise. The issue of feature selection is resolved by deep learning (DL) approaches. One component of ML is DL, and DL can automatically extract key features from unprocessed input data. Cognitive and information theories served as the foundation for the development of DL algorithms.[11]

7.7 AI'S PRIMARY USES IN THE COVID-19 EPIDEMIC

7.7.1 Early Infection Recognition and Diagnosis

AI is able to quickly spot abnormal changes and other warning signs that might alarm both individuals and healthcare authorities. Faster, more economical decision-making is facilitated by it. It helps in the creation of a cutting-edge diagnosis and improved performance for the Covid-19 cases.[11]

7.7.2 Treatment Supervision

AI can create a smart platform for the virus's spread. It is also possible to create a neural network to extract the visual signs of this illness, which would aid in the correct monitoring and care of those who are afflicted. It is able to provide daily updates on the patients' conditions as well as guidelines for dealing with the Covid-19 epidemic.

7.7.3 Locating People's Contacts

By locating clusters and "hot spots" and analysing the virus's level of infection, AI may successfully track contacts for individuals and keep an eye on them. It can forecast the progression of this illness and its likelihood of recurrence.

7.7.4 Case and Mortality Projections

Using information about the dangers of infection from social media and other media platforms, this system can monitor and predict the nature of the virus and the likely extent of it. It can also forecast the number of positive cases and fatalities in any area.

7.7.5 Drug and Vaccine Development

AI is applied to analysing Covid-19 for pharmaceutical research data that is now accessible. Designing and developing drug delivery systems can benefit from it. When compared to traditional testing, which is a hard process for a human to do, it is vastly sped up by this technology. It could help in the search for potent drugs to treat Covid-19 victims. It has grown into a powerful tool for creating screening tests and vaccines. AI supports clinical trials throughout the creation of the vaccine and speeds up the process of producing vaccines and medicines.

7.7.6 Lessening the Complexity of Healthcare Work Professionals

Healthcare workers have a particularly heavy workload due to the increase in the cases. In this case, AI is employed to lighten the workload of healthcare professionals. It provides information to professionals regarding this new disease and aids techniques. AI can improve patient care in the future and handle more possible problems, which will lighten the pressure on doctors.

7.7.7 Disease Prevention

Using AI as a guide can provide current information that is useful for the prevention of future viruses and disorders. It pinpoints characteristics, root causes, and mechanisms behind illness spread. This technology will be crucial in the future to combating subsequent pandemics and outbreaks. It can fight many different ailments and act as a preventive measure. Future healthcare will be more predictive and preventive thanks in large part to AI.[11]

7.8 CONCLUSION

Nowadays, many medical and healthcare devices, such as smartphone applications, portable biometric sensors, and genomics, give a lot of data. However, we need to understand how to use this information. Studies based on these facts, for instance, might provide fresh viewpoints on how to improve healthcare through technical, clinical, and ethical approaches. An examination of these medical practises suggests that patient-specific medical specialties or customised medicine are currently being used to their full potential. A more accurate prognostic framework is constantly being developed thanks to the new technologies that are helping us deal with these things. The businesses offering clinical transformation and healthcare analytics services do help to provide better and more productive results. Reduced analytics costs and the detection and prevention of data-related fraud are all common objectives of these businesses. The handling, sharing, and security of private data are among the federal concerns that almost everyone has to contend with. Better understanding, diagnosis, and treatment of numerous diseases have been made possible by the pooled data from medical institutions. Additionally, this has aided in laying a stronger, more effective foundation for individualised healthcare. The current healthcare sector has adopted data science analytics in clinical and healthcare procedures as a result of realising the possibilities of big data. Quantum computers and supercomputers are assisting in considerably lowing the time it takes to extract valuable information from massive data.

REFERENCES

1. Ozdemir, S. (2016). Book: Principles of data science. Packt Publishing, Birmingham-Mumbai.
2. Nilsson, N. J. (2014). *Principles of artificial intelligence.* 1st edn., eds. Morgan Kaufmann., Berlin, Heidelberg: Springer.
3. Haleem, A., & Javaid, M. (2020). Vaishya effects of COVID 19 pandemic in daily life. *Current Medicine Research and Practice.* doi:10.1016/j.cmrp.2020.03.011
4. Bai, H. X., Hsieh, B., Xiong, Z., Halsey, K., Choi, J. W., Tran, T. M., Pan, I., Shi, L. B., Wang, D. C., Mei, J., & Jiang, X. L. (2020). Performance of radiologists in differentiating COVID-19 from viral pneumonia on chest CT. *Radiology.* doi:10.1148/radiol.2020200823
5. Hu, Z., Ge, Q., Jin, L., & Xiong, M. (2020, February 17). Artificial intelligence forecasting of COVID-19 in China. *arXiv preprint* arXiv:2002.07112
6. Russell, S. J., & Norvig, P. (2016). *Artificial intelligence: A modern approach.* Malaysia: Pearson Education Limited.

7. Smeulders, A. W., & Van Ginneken, A. M. (1989, June 1). An analysis of pathology knowledge and decision making for the development of artificial intelligence-based consulting systems. *Analytical and Quantitative Cytology and Histology*, 11(3), 154–165.
8. Gupta, R., & Misra, A. (2020). Contentious issues and evolving concepts in the clinical presentation and management of patients with COVID-19 infection with reference to use of therapeutic and other drugs used in Co-morbid diseases (Hypertension, diabetes etc.) *Diabetes and Metabolic Syndrome: Clinical Research and Reviews*, 14(3), 251–254.
9. Gupta, R., Ghosh, A., Singh, A. K., & Misra, A. (2020). Clinical considerations for patients with diabetes in times of COVID-19 epidemic. *Diabetes and Metabolic Syndrome: Clinical Research and Reviews*, 14(3), 211–212.
10. Mohamed, K. B. N. R., Sharmila Parveen, S., Rajan, J., & Anderson, R. (2021). Six sigma in health-care service: A case study on COVID 19 patients' satisfaction. *International Journal of Lean Six Sigma*, 12(4), 744–761.
11. Puttagunta, M., & Ravi, S. (2021). Medical image analysis based on deep learning approach. *Multimedia Tools and Applications*, 80(16), 24365–24398.

8 Semantic Technologies in Next Era of Industry Revolution

Industry 4.0

Fatmana Şentürk and Musa Milli

CONTENTS

DOI: 10.1201/9781003310792-8

8.1 INDUSTRY 4.0 CONCEPT

The development of technology in recent years has brought many conveniences in our daily lives. Thanks to technological developments, systems that are constantly developing, that can make decisions on their own, and that can implement these decisions have emerged. In addition, with the integration of developing smart devices into these systems, the solution of any difficult problem has become much faster and easier. Such developments of technology have also caused changes in industry concepts.

The concept of Industry as we know it today first emerged in the 1700s. The concept of Industry 1.0, which constitutes the first step in industrial development, started in 1760 and continued until the mid-1800s. In particular, the development of vapor-powered machines and their use in different production areas prepared the ground for mass production.

With the wide use of electricity at the end of the 1800s, electricity began to be used in production areas. The first mass production as we know it today was made. This development led to the end of the use of vapor machines in large factories and the transition to a continuously flowing band-based production system. This period continued until the beginning of the 20th century.

The concept of Industry 3.0, on the other hand, covers the period starting from the 1960s until 2011. The developments in the field of computers and the inclusion of computers in the production processes both accelerated the production process and provided many conveniences. For example, jobs that are dangerous started to be done through robotic devices. However, these devices are not in a structure that can make and implement decisions on their own. The work that these devices will do is defined in advance and fixed.

The concept of Industry 4.0 started in Germany in 2011 with a project of the German government to promote deep computerization and conceptual innovation in manufacturing. In this context, it is aimed to bring information technologies to the foreground. In other words, it is aimed to collect, store, and process information and data on the existing system and to reveal meaningful conclusions. In addition, it is planned that these results will be handled by the system and automatically passed to the next step. Even sharing the data collected and stored in this system is considered as a part of the system. All this development process of the concept of the industry is shown in Figure 8.1.

The active use of sensors, smartphones, computers, and the Internet in almost all areas of daily life has made such an integration a necessity in production processes. Therefore, these technological concepts have started to be used in industrial activities. At every step of production, one or more of these components work in integration with each other, thus speeding up the production activities, preventing an error that may occur during production, or producing a higher quality product. In this way, it is seen how important the concept of Industry 4.0 is.

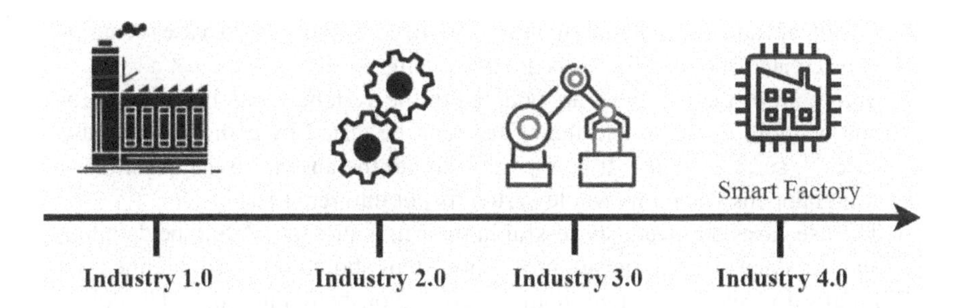

FIGURE 8.1 Evolution of the concept of the industry.

As a result of the increasing global competition environment and companies' desires to use their limited resources such as raw materials, devices, and personnel effectively and efficiently, technological approaches have started to be developed for business/production processes in companies. Especially in areas such as personnel planning, device calibrations and maintenance, evaluation of product quality, and determination of customer satisfaction, the power of information technologies has begun to be utilized. Thus, benefits such as efficient and effective use of resources and fast and on-site decision-making have been provided for companies. In addition, the probability of errors in the products produced in the enterprises that use the automatic production process is lower; therefore, customer satisfaction is at the highest level. With the concept of Industry 4.0, it is ensured that the products produced can quickly respond to the individual needs of the people, and a flexible production system is created.

When companies are contemplating their decisions on the concept of Industry 4.0, it is not enough for some cases that only the machinery and equipment infrastructure is sufficient.[1] For these cases, these machines and equipment must work in harmony with each other and obtain meaningful relationships and results from the data obtained during production. In addition, these inferences should be included in the system in the form of feedback in accordance with the system. In other words, the concept of Industry 4.0 must be integrated into existing systems. The following seven-step procedure model can be applied to achieve this integration:[2]

1. First, experts should be trained to understand the functioning of the system, to be informed about the process, and to ensure that the information is included in the system. At the same time, these experts should exchange ideas with the senior management and the results obtained should be adapted to the system.
2. A comprehensive use case list should be developed for the steps followed in the system. During the design of use cases, parallel works should also be modeled in real time. In addition, a rough cost analysis for these processes should be made.
3. The scope of these created use cases should be constricted to be more specific and with minimum cost and maximum benefit.

4. Customers and raw material suppliers should be included in use case scenarios to be created and therefore to be in the system.
5. Before all of the use cases are adapted to the system, it can be selected for one or more of them, and the system can be tested by using selected use cases to see the results. It is important to get the answer to the question of what happens when this whole system is put into real life.
6. The use cases that were successful in their first attempts should be detailed, and the stages in the relevant production steps should be clearly defined.
7. Finally, the relevant system should be continuously monitored, and its development should be continued through the feedback received. In other words, new use cases created for new situations or use cases that were not included before should be integrated into the system.

With all these application procedures, the aim is to automate the production processes by integrating with digital systems and to provide maximum benefit to the enterprises. In other words, an industrial automation system has emerged with digitalized production processes, and the need for manpower has also decreased. In addition, human errors are minimized.

This chapter will be continued with the application components of Industry 4.0, the algorithms used, the technological components of Industry 4.0, the enrichment of all these application processes and components with ontologies, the elimination of various problems through ontologies, and finally with the advantages and disadvantages.

8.2 IMPLEMENTATION OF INDUSTRY 4.0 STEPS

Businesses integrating Industry 4.0 should apply different steps, such as data collection, data storage, processing, etc. to their systems. The most important feature of Industry 4.0 is that it collects the data collected from the peripheral units in a central system, analyzes this data, and reaches a conclusion. This result can be in the form of a report to the end-users, or it can be as feedback to the system in the enterprise. First, the data is obtained, then the data is stored; various analysis processes are performed on the stored data, and finally, the situation assessment is made according to the analysis results. All these steps are given in Figure 8.2. Details of each step are given later in this section.

8.2.1 Data Collection

The first step for companies to implement the Industry 4.0 concept internally is to obtain data. Data in a company can come from many different sources, such as

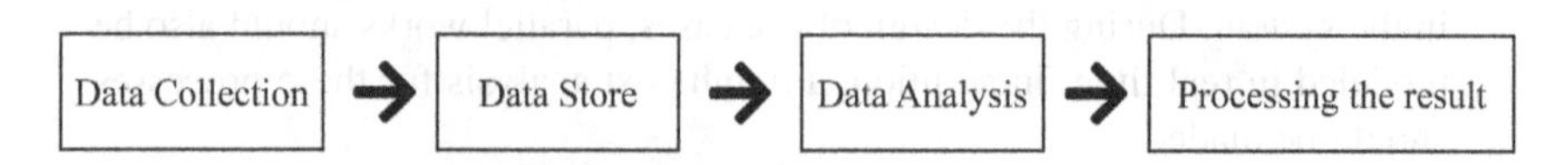

FIGURE 8.2 Implementation steps of Industry 4.0.

cameras, temperature sensors, humidity sensors, and various devices that measure chemical changes. The collected data can be chemical or physical, and it is necessary to transfer these data to somewhere continuously or intermittently. For example, suppose that data is collected from a company that manufactures an electronic device part. Thanks to the sensors, we can learn whether the manufactured part is working correctly or not. First, it can be checked whether the part is produced in standard sizes and in the form of a visual of the part produced using camera systems. In the next step, it can be checked whether the electronic device produces correct values with the help of sensors. Thus, the quality control of the produced parts can be done automatically.

8.2.2 Data Storage

The second step in the Industry 4.0 concept is the storage of the collected data. These data can be transferred to the storage units via a physical layer, or they can be transferred to the storage unit without a physical connection such as Internet, Wi-Fi, or bluetooth. This storage unit can be a server physically located within the company. If a slightly more flexible hierarchy is desired to be established, the relevant data can be also stored in cloud systems. In addition, the correct transmission of this information and its readiness for processing are also handled at this step.

8.2.3 Data Analysis

In today's world, where competition between companies is increasing day by day, companies apply many different methods to improve themselves and gain maximum benefit. For example, a problem in the company's production line can be described as a loss for the company. For this reason, the sooner the company intervenes in any problem, the sooner it has the potential to continue production. Therefore, it is of great importance to constantly listen to the system and to anticipate any negative situation. In addition, sometimes businesses should make an assessment about customer satisfaction in order to develop better quality products. For this reason, it is necessary to analyze and evaluate all kinds of comments from the customer. Other examples can be given for these evaluation methods of the firm. However, on the basis of all these examples, companies need to process the existing data and draw a direction according to the value they obtain. This data processing part constitutes the third step for businesses that adopt the concept of Industry 4.0. Details of these algorithms are given in Section 8.3.

8.2.4 Processing the Obtained Result

Companies can analyze and process the data they obtain, store, or store from different sources. However, these results should also be used. These results can be used in two different ways. The first of these is to enable the system to make automatic decisions and intervene in the system depending on the analysis results obtained as a result of continuous monitoring of the system. The second is to show the results of the analysis obtained to the end-users in a meaningful way. End users can be company senior managers as well as technical personnel. For example, it can give the company managers an idea about the sales of the company in the last year, determine the reasons for the decrease in the production of a product, or give information about the position of the company in the future. Company managers guide a company in making decisions about production planning, product preferences, in short, its future, in line with the analyzed data.

8.3 ALGORITHMS USED WITH INDUSTRY 4.0

Techniques such as machine learning (ML) and data mining are used to detect and predict possible errors in production processes to increase production efficiency and reliability. We can group these techniques into four main headings as data-based techniques, physical model-based techniques, knowledge-based techniques, and hybrid techniques according to the input parameter they use.[3]

8.3.1 Data-Based Techniques

Data-driven techniques are methods that provide significant improvements, especially for smart manufacturing and device maintenance processes. Using the data collected from the devices, the probability of system failure at any time is calculated. In other words, sensors and smart devices can calculate the life of a device for us. The application of these methods to companies enables companies to work uninterruptedly.

We can classify data-based techniques as statistics-based methods, machine learning methods, and deep learning methods. Machine learning and deep learning methods deal with real-world problems and produce successful results. Both techniques have the ability to learn and adapt themselves to new situations. Statistical methods are making probabilistic and mathematical analyzes of existing data. As a result of these analyzes, it classifies the data.

8.3.2 Physical Model-Based Techniques

Physical model-based techniques model the life of a device or the occurrence of a physical failure using mathematical and physics methods. For example, for a continuously operating machine arm, factors such as the coefficient of friction, the angle of rotation of the arm, and the maximum load that the machine arm can carry are the parameters used in estimating the term of this machine arm. If this machine arm can be modeled in a computer environment with all its parameters, it can be predicted when this machine arm will fail.

8.3.3 Knowledge-Based Techniques

Knowledge-based techniques keep the attributes of a computational model in the form of a domain and operate reasoning mechanisms by processing these symbols.[3] These techniques measure the similarity between a new situation and predefined situations and make the most appropriate decision based on this similarity rate.

Knowledge-based techniques can also use knowledge graphs, ontologies, rule-based systems, and fuzzy systems.

8.3.4 Hybrid Techniques

Hybrid techniques are techniques created by taking the strengths of two or more techniques. For example, a combination of both physical models in which physical parameters are evaluated and a knowledge-based technique in which the results of these parametric calculations are interpreted can be used to predict when a device will fail. Thus, the results closest to reality can be measured.

However, due to the heterogeneous nature of the data coming from the system, the information obtained from these data is also in a complex structure.

8.4 TECHNOLOGY COMPONENTS OF INDUSTRY 4.0

Industry 4.0 application processes include many technological components such as the Internet of Things (IoT), cloud computing, data analytics, augmented reality and virtual reality, robotics, and the semantic web, by integrating all these technology steps.[4] In this section, these technological components are examined.

8.4.1 Internet of Things

Internet of Things (IoT) is a network where signals from all kinds of sensors, cameras, detectors, etc. are connected with each other.[5] IoT systems are capable of real-time detection and processing by receiving information from all kinds of technological devices. In businesses implementing Industry 4.0, this technological component can be used for any error due diligence. For example, the existence of such a sensor system in a company producing frozen food can be used to detect possible temperature increases.

8.4.2 Cloud Computing

Cloud computing is a service-based architecture developed for computers and other smart devices with Internet access that provide access to the desired data at any time, scalable, with storage and computing facilities. Cloud-based systems can share the data they store with other systems or use the resources of other systems for calculations, thanks to their service-based architecture. These systems are very important for firms using digital systems. For example, all stages of the production line in a factory can be transferred to the cloud architecture so that the system can be monitored from anywhere at any time.

8.4.3 Data Analytics

Data analytics is a method that enables the processing of raw data to extract meaningful information and to find potential trends of products/customers. Businesses are used data analytics techniques to make more informed decisions to improve themselves. For example, in commercial firms, data analytics are widely used for different purposes such as increasing the market share of firms, modeling prospective customer attitudes toward a product, and estimating the life span of an electronic component.[6]

8.4.4 Augmented Reality and Virtual Reality

Augmented reality is the creation of a virtual copy of the real-life universe by adding data such as graphics, sound, and images through a computer. Virtual reality, on the other hand, is that this created virtual world is perceived in three dimensions using various devices instead of being perceived on the computer screen, and it includes interacting with different senses such as hearing, smell, etc. These two concepts allow simulating possible situations that have not yet occurred in real life. It is a very useful technology for simulating the user experiences of manufactured products, especially for businesses. For example, an automobile company can use these technologies to measure driver reactions for its new vehicle by using various inputs such as different weather conditions and speed limits. Thus, if there is any condition that drivers may have difficulty with, it is determined, and this problem can be arranged before vehicles are produced.

8.4.5 Robotics

With the use of continuously flowing band-based production systems in the industrial field, production activities have accelerated. However, with the desire to increase this production process over time, robots, robot arms, and robotic devices have started to be used. Thus, a sensitive, flexible, and fast system has emerged by using minimum manpower. For example, using manpower for soldering operations in a microchip factory can cause both errors and slow production. The use of robot arms for such a system ensures that the margin of error is reduced.

8.4.6 Cyber-Physical Systems

Cyber-physical systems correspond to the interaction of physical mechanisms and software systems. This interaction can be the continuous monitoring of any system, or it can be a signaling system used to change the status, location, or operation of a device in the system. Cyber-physical systems are one of the most fundamental technological components for Industry 4.0, considering their ability to monitor and interfere with the system. These systems, which provide machine-human interaction by using computer systems, aim to accelerate production processes and increase efficiency and quality. These systems create an automated production process by using embedded software or working with software's intelligent decision-making systems.

8.4.7 Semantic Web

The semantic web is a technology that emerged as an extension of the existing web and enables the conversion of standard data into a form that can be understood by machines. Semantic web technologies provide the link between data that is not of a standard type. For example, it ensures that images from a camera and different types of data from sensors such as heat, temperature, and humidity are within a specific range and standard. In other words, communication between heterogeneous data can be provided through the semantic web.

8.5 INDUSTRY 4.0 AND ONTOLOGY

Today there are various types of data such as pictures, sound, video, text, etc. These data should be processed with the other types by the information systems. How can we provide this interoperability? The answer is the semantic web technologies.

8.5.1 Definition of Ontology

The semantic web technology is a structure with the purpose to interpret the data by computers like a human. This technology can develop algorithms that recognize people's requests and respond accordingly.[7] By using the semantic web, information turns into a structure that can be understood and processed not only by humans but also by computers. In order to create this processible and understandable structure, meta models of the data must be defined. This definition is also provided through ontologies.

Ontologies are metadata in which the concepts specific to a domain, the relationships between these concepts, and the instances of the concepts are defined together.[6] The concepts and their relationships in the real world are also represented as semantic relationships between these concepts. As stated in another way, ontologies are structures that help store and represent real-world data in a way that keeps the patterns within that data and their semantic relationship to each other. Ontologies have a various range of uses and lots of ontologies are constituted automatically or semi-automatically for different domains.

Ontology is a way to help with data combination and interoperability from heterogeneous sources. The ontology may be employed for many different purposes like representing data, extracting information, integrating data from various sources, transforming between data types, etc.

Ontologies can store the data in triple formats. Triples consist of objects, objects, and predicates. By using these triple structures, domain-specific rules and restrictions can be defined. In addition, field-specific samples selected from the relevant field can be stored through these triples. These triples showed by special data languages like Resource Description Framework (RDF), RDF Schema (RDFS), and Web Ontology Language (OWL). RDF can be thought of as an extended version of XML that provides meaningful modeling of data to be processed by the computer. XML only has features such as data storage and data manipulation; it cannot store any statement about the meaning of the data. RDFS is a slightly improved version of RDF. It can define the triples in the domain in terms of objects, relations between objects, properties, and values that properties can take. OWL is an ontology definition language developed by the W3C in 2002. OWL includes a more powerful inference mechanism than RDF and RDFS languages. It has three sublanguages, namely

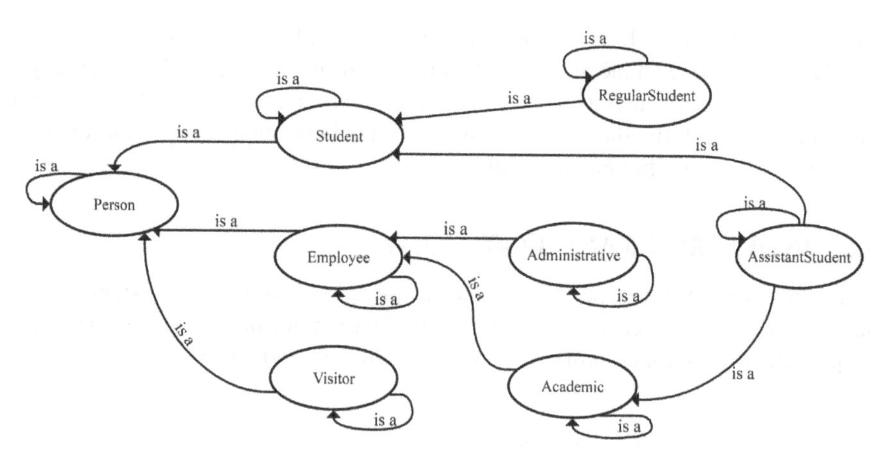

FIGURE 8.3 An example view of person ontology.[8]

OWL, OWL Full, OWL DL, and OWL Lite. An example of the ontology that defines person ontology is shown in Figure 8.3.

Ontologies can store data types obtained from various sources by using metadata of data. These data are structurally and semantically different from each other, so they are difficult to process in information systems. In order for the data from these heterogeneous sources to work together and to obtain a meaningful result, they need to make a link and communicate continuously among themselves. For this purpose, ontologies are integrated into information systems in order to ensure starting related connections and continuity of communications of these links. For example, in a production system, data from sensors such as color, temperature, and humidity can be represented in a suitable format via ontologies. In addition, the obtained data can be used as input data for a different system.

Ontologies have the ability to store rules related to the field in which they are defined in addition to storing information. Information that is not clearly defined in the current ontology can be revealed by using these defined rules in the ontology. In order to obtain this information, special queries should be written and should be operated on the inference mechanism of ontologies. For example, let's say that in an ontology, A is related to B and B is related to C. When we look at this structure, we can understand that A is indirectly related to C. In order to reach the same conclusion on the computers, the inference structure on the ontology is used.

8.5.2 Ontology and Industry 4.0

In Industry 4.0, the ontologies use substeps such as capturing data from a device, converting the data into understandable and machinable form for computers, communicating devices to each other, analyzing the obtained data from the various devices, and explaining the obtained analysis, etc.

Information systems which are used ontology-based approaches can be integrated with Industry 4.0. In these systems, ontologies can be used in any part of

Industry 4.0's application layers. In fact, ontologies are used more for different operations such as collecting data, preprocessing data, storing data, and improving the quality of missing or noisy data, but the ontologies also can be used for making a decision. In short, it is possible to use ontologies at every step where data is available. For example, ontologies can be used to collect data from sensors. Data can be formatted thanks to domain-specific restrictions provided by ontologies. If the data provided from the sensors are planned to be given as input to another system, the input parameter to be sent to the other system can be obtained using ontologies. The detection of corrupt or missing data in the current system is provided through ontologies, and the data contained in the ontology can be used so that this deficiency does not affect the system.

In the step of decision making, ontologies can be used to diminish the search space. By using the inference mechanism provided by ontologies and rule-based data mining algorithms, it can become easier for the system to make decisions. For example, ontologies can be used to evaluate customer feedback about a product. The comments about the product received from the customers are separated and the unnecessary parts are removed and then processed. At this stage, the synonyms of the words used in the comments can be included in the system through ontologies. Thus, even words that have not been encountered before can be used in product evaluation. As a result, the features, pros, and cons of the products that need to be developed can be evaluated from the comments received from the customers. In addition, even if the comment of a product is written in different languages, thanks to the language support of ontologies, comments can be evaluated. The integration of any ontology into one of the relevant steps in Industry 4.0 is given in Figure 8.4. This integration can be in all the steps, or it can take place in only one step.

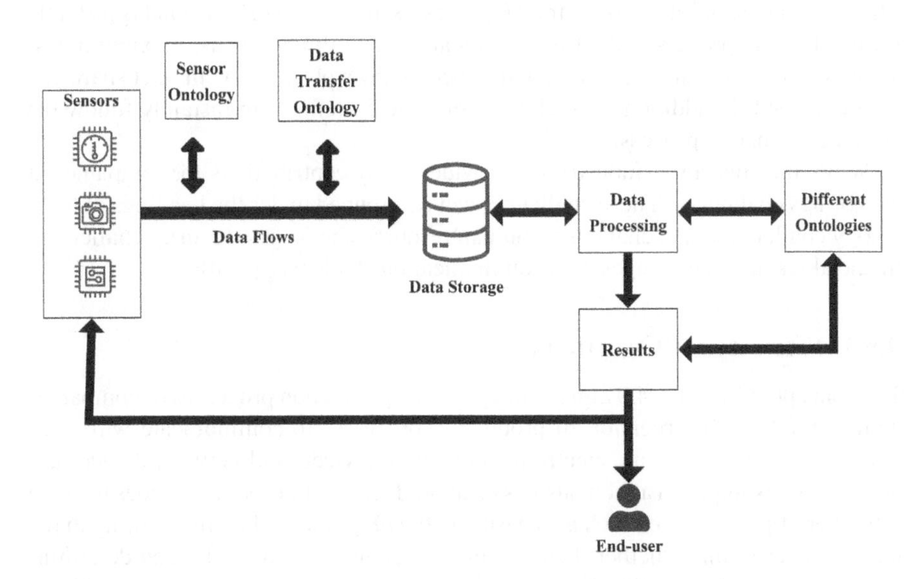

FIGURE 8.4 A general view of the integration of ontology in Industry 4.0.

8.6 THE CONVENIENCES AND DIFFICULTIES IN THE IMPLEMENTATION PROCESS OF INDUSTRY 4.0

In the companies where Industry 4.0 is applied, the existence of technology can be mentioned in the process from the entry of the raw material to the production of the product and even the delivery of the product to the end-user. In this whole process, many sub-units such as factory, supplier, logistics, and customers establish a communication network within themselves. Real-time sensors, smart devices, and different technological components are used in all areas of this network, and all these components need to be automatically communicated among themselves. In addition, different parameters should be optimized because of instant work intensity, physical capacities, etc. from time to time in this communication network. So, it is necessary to develop systems that can automatically share the incoming requests and transfer the results to the relevant units.[9]

Industry 4.0 is an important strategy that increases the competitiveness of companies, thanks to its advantages such as increasing the quality of production services and including flexible and transparent production processes. In addition, companies adopt the concept of Industry 4.0 to overcome difficulties such as increasing customer satisfaction, producing personalized products, optimizing use of raw materials used in production, and keeping the logistics processes of products under control. In addition, companies aim for the devices working in their existing systems to work optimally with each other. In this direction, approaches such as self-inspection of devices and giving shutdown or stop commands to the system when necessary in order to prevent a possible error or breakdown in production are included.

With the Industry 4.0 approach, customer preferences and ideas can be incorporated into the production processes. For example, when a customer orders a personalized item, they may request a color or size change while the process is ongoing. Thanks to the flexibility of Industry 4.0 processes, the customer's demand is instantly included in the process and customer demands are fulfilled. Thus, maximum customer satisfaction can be ensured, and the company's place in the market share can be guaranteed. In addition, in such a system, the customer can instantly follow the product production process.

While the concept of Industry 4.0 provides many contributions for companies, it also brings challenges. These challenges can be grouped under the headings of technology challenges, data challenges, human resource challenges, security challenges, financial resource challenges, and communication challenges.[4, 10]

8.6.1 Technology Challenges

The concept of Industry 4.0 aims to integrate the production processes of companies with technology. In order for all production processes to communicate with each other, a number of sensors, electronic and smart devices, and network devices are needed. For example, in a food business that produces dairy products, it goes through a number of preprocesses such as measuring the pH value of the milk coming to the facility, determining whether there are microorganisms in the milk, and determining the fat ratio of the milk. Depending on the type of product to be produced from milk, the milk must go through different processing steps. In a company whose

Industry 4.0 integration has been completed, data must be collected and processed automatically in all processes from raw material input to the production of the final product, and the environment in the process must be present before the product moves to the next step. The automatic preparation of these processes through the system will speed up the production process. However, for some businesses, this integration may not be fully achieved. In our example, pH measurement and determination of the fat ratio can be determined automatically in the dairy enterprise, while the detection of some microorganisms is done by human power because a device that can automatically detect some microorganisms has not been produced yet. Existing technological deficiencies can be found not only in dairy businesses but also in different operations in different businesses.

8.6.2 Data Challenges

The fact that companies always have devices where we can collect data is not enough to adopt the concept of Industry 4.0. Data from these devices must be within a certain range and standard. For example, different parameters such as bandwidth and flow rate of the data coming from a camera must be in specific standards. Similarly, it is necessary to determine the types of data to be received from the sensors, determine the value range, and determine whether there is noise in the incoming data. At the same time, different problems such as where the data will be stored, data cleaning, and preprocessing steps can be evaluated in this step.

8.6.3 Human Resource Challenges

In the concept of Industry 4.0, the human resource component can be considered in the sense of performing repetitive and routine works through automation and robots and using the existing manpower for jobs that require qualifications and skills.

It may lead to the emergence of new job definitions for new concepts to be formed with Industry 4.0. In other words, while a mechanical master is sufficient for businesses that previously used a mechanical device, a master who understands robotic devices is needed with the transition to computerized and autonomous devices. At this stage, new business lines should not be considered only for the repair of machines. There is a need for engineering fields that will design and develop these devices, communication specialists to ensure communication in these areas, as well as computer specialists who take part in processes such as data processing and data and network security. In order to meet this need, it is necessary to train personnel with a high level of education, knowledge, and skills. This process is costly and takes a long time.

8.6.4 Security Challenges

The lowest stage of the Industry 4.0 process is based on data and data communication. The safe and correct transmission of data to storage units and their storage are of great importance for companies to manage their processes. Therefore, it is necessary to transfer the data collected from the sensors without changing them or to take precautions against theft in the areas where they are stored. In addition, it should be prevented other devices from infiltrating of the data transmission network and receiving information during any transfer phase. This data communication in the internal network of the companies stores a lot of data for the company, and the fact that it is in the hands of third parties directly affects the commercial future of the company. For example, let's say that there is a company that produces fabric and makes automatic printing. Let this company include the concept of Industry 4.0 in all its processes. If there is any leakage from the outside to the printing device, it may cause the pattern to be printed on the fabric to be changed. This change may cause the company to have problems with customers and damage its commercial reputation. In addition, it may not be enough to just close a security vulnerability related to the products produced. In such a system, it is important to protect the commercial records and logistics details of the companies.

8.6.5 Financial Resource Challenges

The concept of Industry 4.0 includes many sub-parameters in terms of development, technology, and education. At the forefront of these parameters are the technological components. It is necessary to purchase technology-compatible devices to be used in the infrastructures of companies, to provide network devices that are the means of interaction of these devices with each other, to store the data received from the system, and to purchase related devices or services. All these components are a financial cost for companies. In addition, it is not enough to install the system once. Ensuring the continuity of this system and having the personnel to take part in all these processes are also cost factors. Therefore, companies need to make serious investments in this field. The inadequacy of government support, the high amount of money spent by companies in this process, and the fact that the process takes a long time reduce the tendency of companies to invest in the concept of Industry 4.0.

8.6.6 Communication Challenges

The most fundamental part of an effective and efficient production in a business is to provide accurate communication by sharing data in real-time between all units involved in production. This communication network consists of sensors, actuators, and network devices working synchronously with each other. However, all these devices may not always be in good and continuous communication with each other. The sensors may be corrupted or there may be external factors that cause the signals produced by the sensors to deteriorate. There may also be domain-related problems between network devices. Different solutions can be offered, such as eliminating the external factors that disrupt this communication line, checking whether the signals produced by the sensors are constantly corrupted, and controlling the network devices.

8.7 DISCUSSION AND CONCLUSION

The concept of Industry 4.0 is the use of sensors, smart devices, and computers in the production processes of companies and the increase in the digital traceability of all these processes. Industry 4.0 can be thought of as a mindset that companies can adapt to themselves, rather than a standard or necessity.

The concept of Industry 4.0 enables companies to monitor their business processes in a transparent and flexible way in real-time by adapting the digital world to them. It allows the requests from the customer to be displayed on the systems instantly and to respond quickly and easily in case of any change. Most importantly, waste of time and raw materials is avoided by using existing resources effectively and efficiently. In addition, it ensures that the dangerous work is carried out with the help of machines and that the situations that may pose a risk to human life are reduced. In addition to all these, it optimizes different parameters such as people, raw materials, machinery, and energy, thus reducing costs for the company.

However, although the concept of Industry 4.0 has all these positive effects, it is very difficult to integrate this concept into companies. There are many different application problems such as evaluating the status of the existing digital infrastructures of the companies, identifying the missing and necessary devices, making all financial analyzes, and taking security measures. In addition, in all these processes, it is necessary to ensure data communication and data standards, complete missing data, or correct erroneous data.

Ontologies can be used to overcome difficulties in data-related stages. In this chapter, an architecture was proposed and brief information was given about the use of ontology in processes.

In order to adapt the concept of Industry 4.0 to companies, the areas that need to be developed are also mentioned. In particular, the existence of new devices that need to be developed in this area and the training of personnel with sufficient infrastructure for the installation and maintenance of these devices is of great importance. With the development of the Industry 4.0 concept, it is inevitable that new business areas will emerge in the future, where different disciplines work with each other. When all these processes are completed, smart process management that can manage themselves and make immediate interventions are targeted. Thus, it is foreseen that smart factories that can use the available raw materials in the best way will be established.

REFERENCES

1. Derya, H. (2018). Endüstri devrimleri ve endüstri 4.0. *GÜ İslahiye İİBF Uluslararası E-Dergi*, 2(2), 1–20.
2. Bildstein, A., & Seidelmann, J. (2014). Industrie 4.0-readiness: Migration zur Industrie 4.0-Fertigung. In *Industrie 4.0 in Produktion, Automatisierung und Logistik*. Wiesbaden: Springer Vieweg, pp. 581–597.
3. Cao, Q., Zanni-Merk, C., Samet, A., Reich, C., de Beuvron, F. D. B., Beckmann, A., & Giannetti, C. (2022). KSPMI: A knowledge-based system for predictive maintenance in industry 4.0. *Robotics and Computer-Integrated Manufacturing*, 74, 102281.
4. Rikalovic, A., Suzic, N., Bajic, B., & Piuri, V. (2021). Industry 4.0 implementation challenges and opportunities: A technological perspective. *IEEE Systems Journal*, 16(2), 2797–2810.
5. Atzori, L., Iera, A., & Morabito, G. (2010). The internet of things: A survey. *Computer Networks*, 54(15), 2787–2805.

6. Şentürk, F. (2021). Ontology for data analytics. In *Smart connected world*. Cham: Springer, pp. 107–123.
7. Berners-Lee, T., & Fischetti, M. (2000). *Weaving the web: The original design and ultimate destiny of the world wide Web by its inventor*. New York, NY: Harper San Francisco.
8. Bravo, M., Reyes-Ortiz, J. A., Cruz-Ruiz, I., Gutiérrez-Rosales, A., & Padilla-Cuevas, J. (2018). Ontology for academic context reasoning. *Procedia Computer Science*, 141, 175–182.
9. İlhan, İ. (2019). Tekstil üretim süreçleri açısından endüstri 4.0 kavramı. *Pamukkale Üniversitesi Mühendislik Bilimleri Dergisi*, 25(7), 810–823.
10. Bajic, B., Rikalovic, A., Suzic, N., & Piuri, V. (2020). Industry 4.0 implementation challenges and opportunities: A managerial perspective. *IEEE Systems Journal*, 15(1), 546–559.

9 Semantic Interoperability Framework and its Application in Agriculture

Ujwala Bharambe, Chhaya Narvekar, Suryakant Sawant and Sulochana Devi

CONTENTS

9.1 INTRODUCTION

High volume of data is produced in many domains as a result of the rapid advancement of modern technologies including Internet of Things (IoT), big data, artificial intelligence (AI), cloud computing, and aerial imagery. With the IoT, actuators,

DOI: 10.1201/9781003310792-9

and sensors can be accessed, managed, and integrated digitally via the cloud, changing the traditional paradigm in terms of how they are accessed and managed as well as translating information into the digital world.[1] This change has allowed new management methods to be developed and improved over traditional methods.

López-Morales et al. (2020) state that this transformation can result in the following benefits: (i) Problem detection: several problems can be identified through monitoring and analysis, including crop status estimation and crop problems recognition (improper water stress, fertiliser use, pest monitoring, fickle weather conditions, and others) by using new technologies, such as sensors, satellite images, drones, various algorithms, mobile applications, and GPS technology before these factors begin to interfere with crop performance; (ii) Productivity improvement:[1] by analyzing and automating processes, overall production can be increased; (iii) Improvement in decision making: to ensure better decisions, it is possible to monitor and analyse various parameters, such as crop water needs, rainfall forecasts, and irrigation schedules, to determine which areas of the land require more water; and (iv) Pattern Analysis: this technique identifies the most cost-effective actions based on patterns, such as crop yields, energy efficiency, and agricultural practices, along with those that should be altered or eliminated.

However, in order to interact, i.e. to access and share this data, the application and data sources must be able to find each other and to agree on the syntax and semantics of data. Interoperability is therefore essential. "Interoperability" is defined by International organisation for standardisation (ISO) as the potential to exchange information, execute programs, or send records among different functional units with the user having little or no need to understand their characteristics. Interoperability is, however, hindered by heterogeneity. The following characteristics of data make it difficult to address this obstacle: (i) Complex: acquired data can be extremely complex in nature because it is represented in different data models and stored in multiple types of formats (such as images, shape files, etc.); (ii) Dynamic: a lot of data is dynamic and constantly changing; (iii) Large: large amounts of data (e.g., satellite data); and (iv) No standard algebra defined: there is no standard set of mathematical operators for merging and integrating different types of data.

It is a significant challenge for application developers to interoperate generated data with the above characteristics. There are no well-defined standards or existing tools to address the semantic interoperability (SI) problem of big data applications. In terms of semantic interoperability, ontologies are among the most interesting topics that have already proven to be the foundation for a wide range of applications. Formal frameworks for analysing, integrating, and communicating data, ontologies facilitate understanding of human language, which enables data to be exchanged and shared between devices and systems, resulting in interoperability. Interoperability is discussed in this chapter in various forms and dimensions. Moreover, we discuss the need for semantic interoperability and how to achieve semantic interoperability via an ontology-based information system in agriculture.

9.2 INTEROPERABILITY AND STANDARDISATION

Having access to information systems that are interoperable has been a problem since 1988, and possibly even earlier. The literature defines interoperability differently. *The Oxford Dictionary* describes interoperability as "the capability to work together". Two interoperable systems can communicate with each other and utilise each other's capabilities. Institute of Electrical and Electronics Engineers (IEEE) has defined interoperability as "the ability of two or more systems or components to exchange information and to use the information that has been exchanged".[2] The International Organization for Standardization (ISO) has defined interoperability as "the capability to communicate, execute programs, or transfer data among various functional units in a manner that requires the user to have little or no knowledge of the unique characteristics of those units".

9.2.1 Types of Interoperability

Layered models can also be used to predict the ability of two systems to interoperate. As an example, Tolk and Muguira (2003) offer a structure of six levels, which include: no connection (no interoperationally between systems), technical (connections between systems and networks), syntactical (the exchange of data), semantic (understanding the meaning of the data), pragmatic/dynamic (compatibility of the information), and conceptual (a common understanding of the world).[3] They proposes a similar six-level model containing connection, communication, semantics, dynamics, behaviour, and conceptual. According to the Tolk's and Muguira's model, these six levels are conceptual, pragmatic/dynamic, semantic, syntactical, and technical.

As shown in Figure 9.1 there are multiple levels of interoperability starting with technical, syntactical, semantic, system and so on.[4]

Technical interoperability is the lowermost level of interoperability (Level 1), which is associated with software components and hardware, platforms, and systems which actuates machine-to-machine communication. This type of interoperability is based on a given set of protocols and a minimum base needed for protocol operation. Two or more systems exhibit syntactic interoperability if they are able to communicate and exchange data.

Specified data formats and communication protocols need to be standardised to achieve syntactic interoperability and extensible markup language (XML) or Structured Query Language (SQL) standards and are among the current tools to achieve the same. This is also true for lower-level data formats such as ASCII or Unicode format in all communicating systems. The prerequisite for achieving syntactic interoperability is that technical interoperability must be in place. However, the prerequisite for achieving syntactical interoperability is that technical interoperability must be in place.

Similarly, syntactic interoperability is a necessary condition for further interoperability. The standards of the Open Geospatial Consortium (OGC) provide a

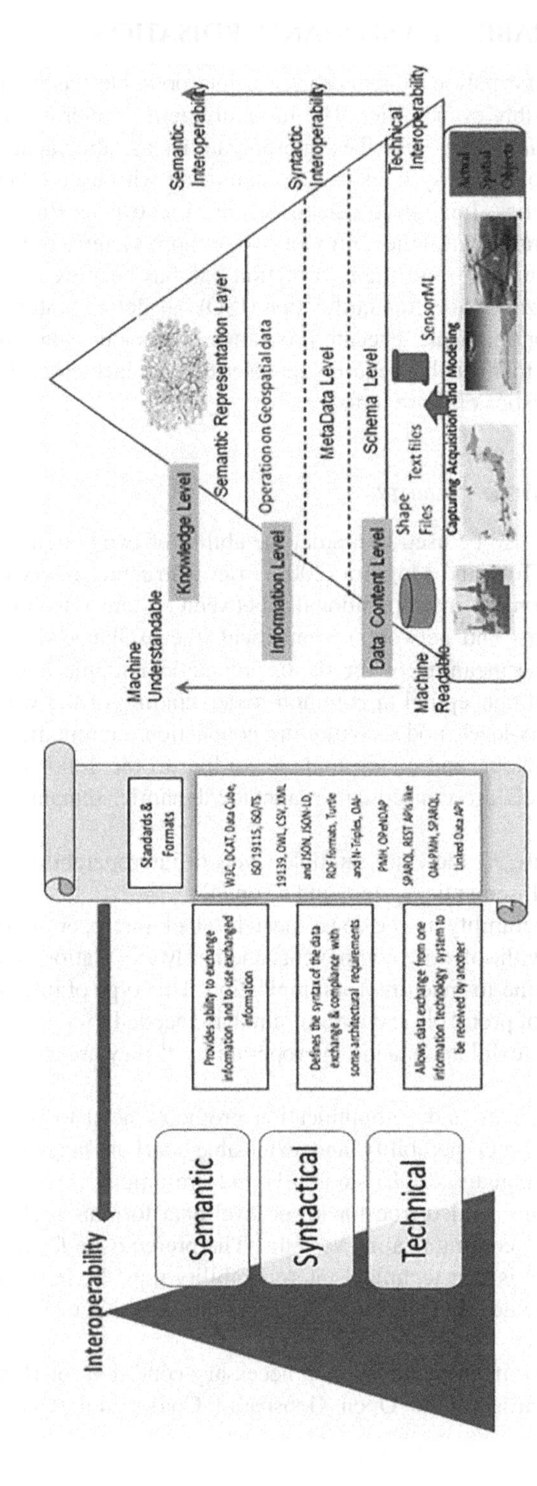

FIGURE 9.1 Levels of interoperability and knowledge discovery.

syntactical basis for data interchange between different user communities for geospatial data. However, this is only an initial step as semantic heterogeneity is still an impediment to full interoperability.[5] In contrast to syntax, which only defines the structure, semantics refer also to the meaning of elements. The heterogeneous nature of data content and user communities (different languages, terminologies, and perspectives) create a semantic problem that existing standards failed to tackle. To achieve interoperability, the system needs seamless access to underlying heterogeneous data sources. In this process, harmonization is required to alleviate the conflicts also known as semantic discrepancy.

Semantic discrepancy is the disparity or heterogeneity present in underlying data sources in its context meaning though they have been represented in the same standardised model and accessed in the standardised way (web services). This semantic discrepancy can be addressed using semantic interoperability referred to as the ability of systems to exchange information that has a common meaning by using agreed standards.

9.2.2 Approaches for Interoperability

There is a considerable amount of research that indicates there are different ways to promote interoperability, like integrating it in application designs at the initial level of the development of devices, but this cannot be easy due to the ownership structure, APIs that permit collaboration, and middleware or ontologies that describe common concepts. Various support methods are shown in Figure 9.2.

(i) *Platform based interoperability:* Data discovery, access, integration, and analysis are among the goals of the interoperability platform. As a result, it has encouraged the adoption of standardised file formats, vocabulary, metadata identifiers, and has worked internationally to achieve its goals.

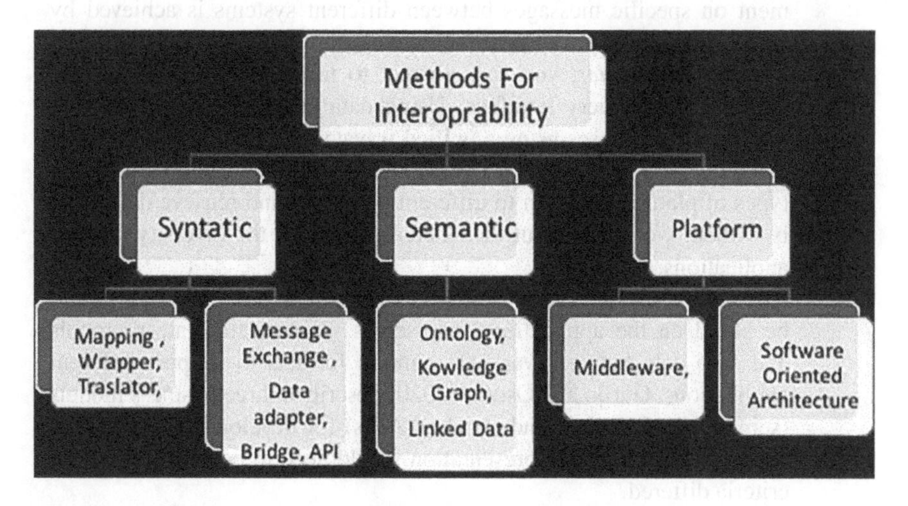

FIGURE 9.2 Support methods for interoperability.

- *Middleware:* The middleware converts the messages between heterogeneous systems into a standard format that all cooperating parties can understand one form to another. One solution to create an intermediary is to use the cloud as an application and service intermediary. Within the cloud, messages can be exchanged between applications or databases can support multiple sources.
- *Service oriented architecture:* Service-oriented architecture (SOA) concepts refer to a way of integrating services across multiple platforms and programming languages to create a complete application. Data exchange can be improved and interoperability between heterogeneous systems is made possible by SOA. Users can retrieve the most current records from a network using specific web services. Different platforms and architectures can communicate and interact with web services in a SOA environment without having to change their design.

(ii) *Syntactic interoperability*: Data interaction between components is called syntactic interoperability. Data formats and structures for information exchange in this area are defined by standards.[6]

The *mapping process* involves transforming data from one format to another to allow the involved systems to work together. For example, an extensible markup language (XML) file can be transformed into Java Script Object Notation (JSON) by mapping at the data level.

- *Wrappers* work to enable interoperability with legacy systems by acting as intermediaries between external systems and legacy systems.
- *Translators* are often used at the application level (between sender and receiver) to translate data from one format to another for the receiver to understand. The translator could be a one-way or two-way translator.
- Collaboration on *message exchanges* occurs at the level of specific data exchanges between different services or applications based on standard data formats understood by both parties, like JSON or XML. An agreement on specific messages between different systems is achieved by categorizing the types of possible messages.
- Using *data adapter*, you can connect to multiple databases using a unified private query interface. Using data adapter, you can connect to multiple databases using a unified private query interface. With this technique, we can create a simple web service that allows different types of platforms to join to different databases and retrieve data. This is without worrying about their heterogeneity or the diversity of their applications.
- Application-to-application *bridges* (better than modification) should be based on the applications themselves. Bridges translate protocols and map data between various common formats to support different applications. Garijo and Osorio (2020) described three business models (semantic web, proxy, and standard APIs) for developing a consensus platform.[7] The data representation would need to be converted if the criteria differed.

- Application programming interfaces (APIs) are methods that allow an application to provide services or data in a form that is readily usable by other applications and systems. Every system and application has APIs that provide public methods that can be invoked by other applications and systems to provide the desired services or data. In general, APIs are suitable only for agreements on specific and common focal points, and a unified standard is necessary. By providing open APIs,[6] the importance and necessity of enabling easy data collection and sharing is emphasised.

(iii) *Semantic interoperability:* As well as data exchange (syntax), semantic interoperability includes the transmission of meaning (semantics).

Through the use of ontologies, knowledge graphs, and linked data, users can discover contextualized, harmonized, and meaningful data that bridges information systems, information sources, and silos by enabling immediate, interactive access to data from all required sources. As a result, customers can respond more quickly to evolving standards, requirements, and new requirements. Integrated data also allows for the preservation of provenance, an essential capability for validating critical decisions. Ontologies provide the formal semantics of a knowledge graph (KG) and linked data. It is conceivable to think of ontologies as the data schema of a graph, while KGs and linked data are instances or actual data.

9.2.3 Need for Semantic Interoperability

Interoperability allows information exchange and processing to be managed more efficiently. In order to ensure interoperability, systems must be able to talk and understand each other's information. In order to achieve this understanding, semantic interoperability is required, as it focuses on the meaning of the underlying information. Having semantic interoperability will make it easier for information systems to share information unambiguously.

Knowledge discovery: Traditional information management system is made up of three layers: data layer, processing layer, and user interface.

Data layer: In the traditional system, data is stored in a database and based on the type of database, the schema gets defined. Data can be visualized in three different layers: data content layer, schema layer, metadata layer.

- *Data content layer*: In this layer, data is stored in different types of spatial databases like Oracle, MongoDB, and MySQL etc. For example, the data for an urban planning department, transport system, traffic data, and road network data each could be stored in different formats.
- *Schema layer*: This layer represents the conceptual model of data. There are two dominant views for schema representation. One is influenced by the physical model of the database and the second one is influenced by the domain data model.
- *Metadata layer*: The layer contains metadata about the underlying data. In metadata, information is described in a structured manner to aid in sorting

and identifying aspects of the information it describes. Schema is a critical part of metadata. A metadata scheme describes the underlying structure of a given data set. A metadata schema explains how metadata is structured, typically addressing common metadata components such as dates, places, and names. Moreover, there are domain-specific schemas, which address specific elements required by a discipline. In the case of metadata, there are many standards. One is provided by the Federal Data Committee while another is by the International Organization of Standards (ISO), and also by the Federal Geographic Data Committee (FDGC) defined Content Standard for Digital Geospatial Metadata (CSDGM). The ISO has established geospatial metadata standard ISO 19115, the e-Government Metadata Standard, e-GMS, etc.

The Processing Layer is responsible for information analysis and query management. The next layer is the user *interface layer* which is responsible for visualization and communication with users as shown in Figure 9.3. Now assuming that the user requests for the transport options available for evacuation during the floods and he/she fires a query through the user interface. This query is again processed by the processing layer, and the data related to the road network in the region is accessed through data source and presented to the user. This appears to be a very simple processing task. However, actually the process of query execution is very complex with many hurdles.

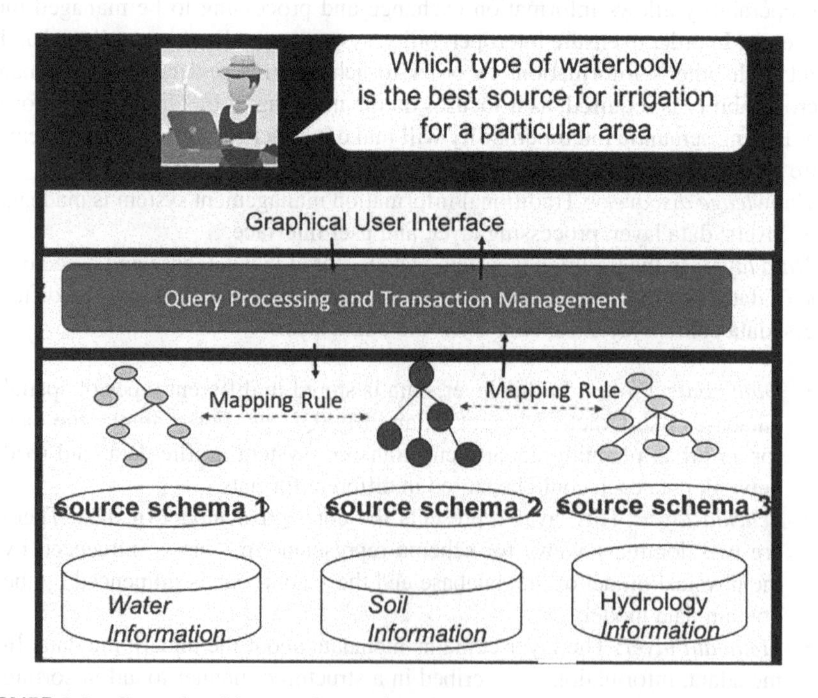

FIGURE 9.3 General architecture for ontology base.

For the query, transportation (road network) data is required. This data available may be in varying formats in different data sources. Also, to generate an efficient flood evacuation plan, there is a need for integrated population data, transportation data (traffic directions and rates, vehicle occupancy, number of vehicles on roads, road network carrying capacities),[8] and destination nodes including evacuation shelters. As the data must come from a variety of different sources with each storing data in a different format, the challenge lies in the integration process to achieve interoperability. Many data collection agencies limit their data collection to particular regions that differ from system to system and are subject to change over time. Therefore, data sources could differ in their data models and coverage. Incompatible, missing, or undefined data for different regions and from different sources may be available. Data may be available for points when it is actually needed for continuous areas. Occasionally, data may not be available for a particular region but may be available for another. With regard to all the above issues, data level integration does not suffice, and semantic level integration is required. This can be accomplished by using ontologies.

9.3 SEMANTIC INTEROPERABILITY

Semantic interoperability is an expertise of the general definition that refers to understanding and interpreting data that is exchanged for the benefit of multiple actors in a system. Standard understanding of data is provided by using common data formats and standard nomenclatures. It focuses on the meaning of the information, not its packaging (i.e. syntax). The following should be included in an interoperable system: data must be clear to be received, processed, and transmitted; an exchange of content between two systems (machine or human) without distortion or delay; a distributed system's ability to exchange data and services between its parts; services and information can be exchanged; data and functions can be combined based on their significance; automating workflows and streamlining the flow of information; and system services can be accepted from other systems and new services can be provided. Through semantic interoperability, organizations can treat data as a living resource that can be reused, combined, exchanged, and understood by a variety of systems.

The development of semantic data repositories with semantic interoperability will enhance analytical capabilities, ease the process of integration of diverse data sources, and, therefore, facilitate the formation of data-centric ecosystems. In addition to these benefits, we can integrate data from a variety of sources and integrate it across multiple consumers; convert the data into a machine-readable format; analyze the data to find patterns and relationships; resolve data discrepancies faster and break down data silos; obtain a holistic view of an enterprise's information landscape (data federation); improving database queries for both efficiency and quality; and improving data discoverability.

Recently, ontology is the tool being proposed and under research for semantic level integration. Knowledge representation is achieved by creating ontologies. With ontologies, these discrepancies can be resolved semantically rather than syntactically. Communication between different organizations can be improved using it as a common language. In order to achieve semantic interoperability, ontologies can be used.[9]

- Provide domain models to facilitate knowledge elicitation and better understanding of the domain.
- Provide a roadmap to understanding fine-grained reusable models and establish a common context
- Provide a common context between different application systems to facilitate interoperability. Facilitate communication among people and agencies by providing a common language and understanding of the structures involved.

9.3.1 Topics Related to Semantic Technology

Semantic technologies are primarily envisioned to increase the usefulness of the World Wide Web by enabling meaningful content to be structured, so that computers and people can work together more effectively. Semantic web has been gaining attention since the early 2000s, and we can categorize it into three overlapping phases driven by a key concept each; that is, this field has shifted its main focus twice during the reconstruction process. The first phase, which spanned the early to mid-2000s, was driven by ontologies. The second phase, which was driven by linked data, spans into the early 2010s. The third phase was and is still driven by knowledge graphs. The following section describes all three main semantic technologies.

Semantics is a broad area in which integration of the information is a fundamental step. Semantic theory also addresses the meaning attached to data elements and their relationship with each other. It is a keystone for understanding the implicit meaning of the data. There is an increasing requirement of information knowledge modelling, integration, management, and reuse. A lot of research is ongoing in the information community to develop and deploy sharable and reusable models known as ontologies.

9.3.1.1 Ontologies

The concept of ontology has travelled a long way from philosophy to knowledge representation. "Ontology" means the study of what is existing. The word comes from classical Greek. Ontologies are applied in many other fields to satisfy different requirements. It was independently determined that knowledge representation is essential for the evolution of artificial intelligence, software engineering, and database communities.[10]

Gruber formally defined ontology in 1993 [11] as an "explicit specification of a conceptualization". Conceptualization is the process of defining and clarifying the concepts of underlying domains. Human minds form ideas through conceptualization. Observable and essential characteristics of the elements are represented mentally (based on human perception). Domain knowledge is formed by explaining these characteristics.

An ontology represents knowledge in three ways: subject, object, and predicate. With modern computers and the Internet, ontology has preserved its meaning but gained a practical aspect. Community of people with different cultural backgrounds who know and understand each other helps to establish a common understanding of a topic. Today, the concept is increasingly related to the World Wide Web and semantic web. A semantic web ontology is a standardized vocabulary that can be shared and transferred over the Internet. Ontologies are a way to represent knowledge and correlations between different pieces of information.

9.3.1.2 Life Cycle of Ontology in Information System

Ontology engineering is still at its maturing stage. Literature offers some ontology engineering methods like the "Evaluation Method" proposed by Grüninger (2009);[12] the "Methontology Method" proposed by López et al. (1999) at an artificial intelligence lab at the Madrid Institute of Technology,[13] etc. In this chapter we are using the ontology engineering method by Sure, Y., Staab et al. (2009) [37]. Ontology engineering is divided into two phases as shown in Figure 9.4. Phase 1 is ontology modelling, and phase 2 is ontology mapping and merging and ontology storage and querying. A detailed description of the ontology engineering phases follows.

9.3.1.3 Ontology Modelling

Generating ontologies is part of the ontology modelling process. There are two ways to use ontologies in systems of information: single shared ontology approach or multiple ontology approach. As it is very difficult to agree on a consensus amongst different agencies when modelling in a single ontology, using a single shared ontology has a lot of practical issues. Therefore, multiple ontologies are commonly used. Ontologies of various types are used for representing data, domain knowledge, application knowledge, procedural knowledge (processes and tasks), and organizational knowledge, among others. The process of building an ontology is challenging, and no standard method exists. It is the ontology modelling phase where the ontologies are constructed for representation of knowledge and vocabulary of domain which can be queried. The ontology modelling process is shown in Figure 9.5. There are two steps in ontology modelling: ontology capture and ontology coding.

9.3.1.3.1 Ontology Capture

Identifying actors and their requirements is the first step in this phase. In the flood management system, nine agencies are involved, and there are several domain

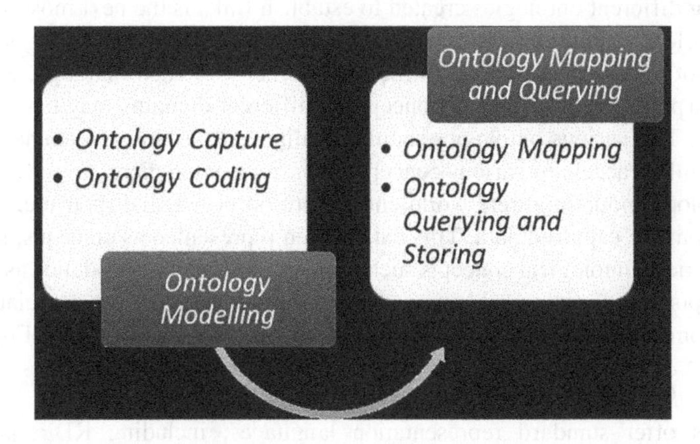

FIGURE 9.4 Life cycle of ontology in information system.

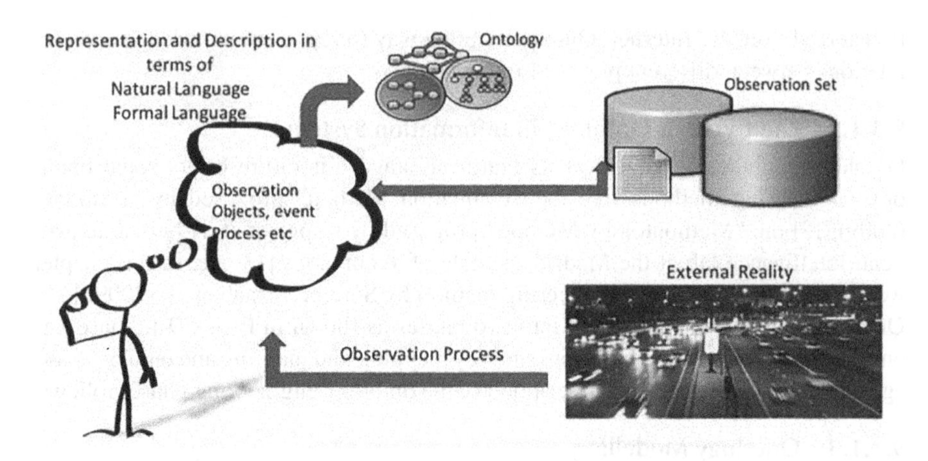

FIGURE 9.5 Ontology modelling process.

ontologies, static data ontologies, dynamic data ontologies, tasks ontologies, and application ontologies. Nevertheless, when designing an application, it is also necessary to define the scope of ontologies.

Following that, it is necessary to identify key concepts within the defined scope of the particular ontology.

The next step is to identify the relationships among the various concepts extracted. There are two types of relations that can be modelled in an ontology: hierarchical and association. Relationships between super classes and subclasses are also called IS-A relationships. Association relationships are modelled as object properties. There are two ways to represent the relationships between spatial-temporal concepts and normal domain concepts hierarchy and association.

Defining data properties is the next step, which binds data values with information. After that, ontologies need to define various restrictions and constraints of the given concepts based on its definitions.

Linking different ontologies created to establish links is the next most important step in ontology capture. Sometimes this can be done explicitly. However, such linkages are not always possible due to the heterogeneity of each ontology, which has its own purpose and granularity. Concepts in different domains may have different definitions. The various ontologies should be aligned to establish a connection that allows seamless access to various concepts in various ontologies.

In ontology modelling, real world entities are observed, and their meanings are derived from the captured data. This data is then represented as concepts, relations, and properties. Ontological concepts such as *River* and *Dam* are modelled as ontological concepts in the ontological language and coded as *owl:class*. An association relationship connects two concepts, as *owl:ObjectProperty* is represented in Figure 9.6.

9.3.1.3.2 Ontology Coding

Ontologies offer standard representation languages, including RDF[1] and RDF schema[13], OWL[2] or Simple Organization System (SKOS). A resource can be

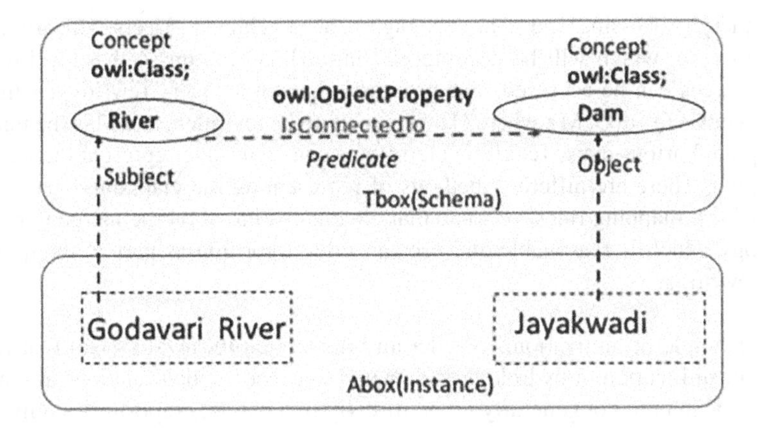

FIGURE 9.6 T-Box (schema level) and A-Box (instance level) representation of ontology.

described using RDF triples composed of a subject, a predicate, and an object. The subject (the resource being described, e.g. Road) is associated with an object (e.g. a junction) through the predicate (the property to be described, e.g. has Road elements or has Proper Part). In RDF, you can type resources and it is constructed by RDF Schema and OWL.[15] OWL enhances RDF Schema to build complex concepts from simple concepts and also adds complex constructors based on description logic. By expressing the interrelationship between subject, object, and predicate as axioms, description logic enables the formal representation of knowledge.

9.3.1.4 Ontology Mapping and Querying

Ontology mapping is used for (i) connecting ontologies with the information sources they describe; (ii) connecting the different ontologies used in the system, and (iii) adding knowledge (query concept expansion) in user query.[16]

- *Ontology mapping:* An ontology mapping technique resolves heterogeneities between ontologies. Heterogeneities include the same term describing different concepts or the same term describing a different concept, as well as different levels of granularity or domain coverage. Identifying the elements in the source ontology that correspond to those in the target ontology is the process of ontology mapping. The mapping process comprises three phases: (i) mapping discovery, (ii) mapping representation, and (iii) mapping exploitation/execution. The goal of ontology mapping is to support instance transformation of heterogeneous data sources.
- *Ontology querying and storing:* Providing unified views to users is the primary objective of interoperability. To extract user information from an ontology based information system, one must query concerns ontologies. An ontology can be queried using SPARQL or RQL. The querying of disparate data sources in an integrated environment can be a challenging task, especially when the data is heterogeneous. On the basis of ontology mappings,

> query processing and data exchange can be achieved. Users will provide query q, which will be decomposed into q1, q2 . . . qn so that local data sources can be accessed. The process is known as query rewriting, which denotes Q'=f(Q, M), where Q is the query to be rewritten, and M is the mapping. Various query rewriting algorithms can be used to represent this mapping. There are different methods of representing the mappings: mapping tables, mapping rules, or some that are created based on the merged ontology which is reasonable and uses an infer algorithm as part of the query rewriting.

As an example of motivation, consider an information retrieval system that makes a way for unification of hydrological data and uncover implicit relations among the features which are not generally found directly in local data repositories. Given the query q1, the concepts are extracted from the query and the query is enriched with added knowledge using geo-ontology matching where the local data ontology and the domain ontology are matched. This will help generate a knowledge attribute set that allows users to retrieve the information. For example, for a user query "Which type of water body is the best source for irrigation for a particular area?" The local data ontology provides only the information (raster data) regarding the presence of different types of water bodies and their attributes (for a particular area). However, to find the best water source for irrigation not only the type of water body is important but also the domain understanding of various of water bodies is important (e.g. the properties of the water content in the water body determines its suitability for irrigation purposes, i.e. pH level <7, sodium content level, Mn and Fe content level < 0.3ppm each, salinity level, hardness <150ppm).

The main purpose of development of ontology concept is to attain integration in order to achieve interoperability. It has been developed to bridge the fields of knowledge sharing and information discovery in the geospatial domain to enable geospatial semantic interoperability.[17] Ontologies are used in the integration process for explicit description of the data source semantics.[18] However, these can differ in the way of integration process employment. Generally, ontology can be employed in three ways: single ontology approach, multiple ontology approach, and hybrid ontology approach.[19]

The adversity of hybrid and multiple ontologies is the lack of a common vocabulary which makes it difficult to compare ontologies from different sources. To overcome this problem, an additional representation formalism defining the inter ontology mapping is needed. This problem of mapping is a well-known problem in knowledge engineering. The mapping identifies semantically corresponding (similar or equivalent) terms in ontologies from different sources.

9.3.2 Linked Data

The year 2006 saw the rise of "linked data" (or "linked open data", if public, free, and open data is the goal). The term "linked data" is generally used to refer to RDF graphs linked in such a way that most of the internationalized resource identifiers (IRIs) in one graph also appear in other, sometimes multiple, graphs. Linked data remained a major driving force for semantic web research and applications and persisted as such well into the early 2010s. It can be understood that a collection of all these linked RDF graphs is one very large RDF graph. Often, big data providers only provide an interface for querying their databases based on SPARQL[3] (referred to as a "SPARQL endpoint").

In the early 2010s, there was an initial discussion about linked data that gave way to a more realistic view. The fact remains, however, that integrating and utilizing linked data turned out to be more difficult than some initially thought. Although interlinks between datasets were initially thought to cure this weakness, it did not seem to happen that way as the shallow non-expressive schemas often used for linked data proved to be a major obstacle to reusability.[20] Data is transparently accessed via hyperlinks in linked data. This helps in integration of data from disparate data silos and connectivity across new and existing enterprise solutions. Despite the huge progress linked data has made along the way, this does not mean that the field or its applications should be devalued. Just having data in a structured format following a prominent standard means that it can be accessed, integrated, and curated with available tools and then made use of. Even so, the search for more efficient approaches to sharing, discovering, integrating, and reusing data is important as ever and is already underway.

The following principles guide the creation and use of linked data: use a hyperlink to reference anything of interest (preferably HTTP URI, since HTTP is ubiquitous and use structured sentences that identify the subject, predicate, and object, utilizing hyperlinks for subjects and predicates and literals for objects. As we use sentences to describe things of interest in the real world, we are creating a powerful web of data, knowledge, and insights that are frequently discovered in a surprising manner.[4]

9.3.3 Knowledge Graph

Knowledge graphs (KGs) represent relationships between real-world instances, such as objects, events, situations, or concepts and illustrate how they relate to each other. The term knowledge graph refers to the fact that this information is typically stored in graph databases and visualized as graph structures.

KGs store the meaning of the data alongside the data in the form of ontologies. In this way, KGs become self-explanatory, allowing data to be discovered and understood in a single place. Inferencing is supported by a semantics that are explicit and includes forms. It is possible to describe the meaning of the main concepts and relationships of KGs through their large volume of items. It may be necessary to adjust data to meet data model requirements according to their recommendations. Furthermore, they allow for the drawing of conclusions and the discovery of new information based on the available data. The KG has also been effective in resolving semantic interoperability conflicts while integrating data from various domains such as human traffic, medicine and agriculture. The main features of KG are

- Organises real-life entities in a graph along with their interrelationships. They are more focused on actual instances (A-Box). Schema (T-Box) plays a minor role.

- Describes the concepts and relations of entities in a schema.
- Provides the possibility of relating arbitrary entities to one another.
- Collects and integrates information into an ontology and applies reasoner.

9.3.4 Ontology-Based Integration and Interoperability

Ontology-based data integration (OBDI) is the data integration technique to integrate different ontologies to capture implied knowledge from heterogeneous data sources to attain semantic interoperability.[2] The standard design requirement of ontology-based information system is to integrate data from non-ontology sources and allow virtual data access; it is divergent from traditional methods such as ETL (i.e., extract, transform, and load).

Figure 9.7 shows a set of criteria adapted from Ekaputra et al. (2017) for performing the data integration (i) language and framework, (ii) data acquisition, (iii) mapping, (iv) transformation, (v) data storage, and (vi) data access.[21]

Language and framework includes approaches such as RDF; OWL; F Logic, e.g. Angele and Gesmann (2006);[22] ML topic maps, e.g. Lee and Kim (2007);[23] and common logics, e.g. Imran and Young (2016) for integration.[24]

Data acquisition methods include ELT; extract, load, and transform (ELT); and ODBA. ETL is used to load transformed data to the target data store; it combines data from different sources and converts this data into a single usable data store.

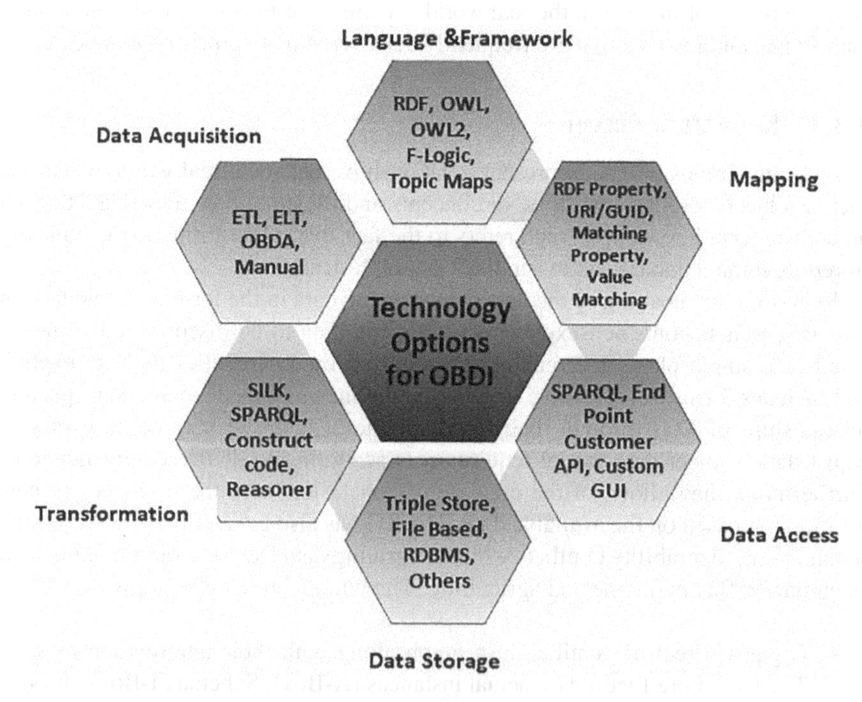

FIGURE 9.7 Technology options for ontology-based data integration (OBDI).

Whereas in ELT raw data is loaded, and transformation takes place to an ontology as required.

OBDA is an approach that allows users to access external data sources by giving users a high-level domain view, convenient vocabulary for rewriting queries, and delegating the queries to relational databases as a virtual RDF graph. Most of the surveyed methods apply RDF property matching and URI and global unique identifier (GUID). URI and GUID are used to link equivalent instances from different ontologies.

With regards to the storage of data, the subcategories of RDF triplestore, file-based, relational databases (RDBMS) are outlined. SPARQL endpoints are the most common alternative for data access followed by customized APIs and GUIs.

9.4 SEMANTIC INTEROPERABILITY FRAMEWORK FOR AGRICULTURE DOMAIN

There are a number of data sources that can be accessed in the agricultural domain, including farmers, sensors, markets, and government data. Using open-data platforms, one can process, integrate, and export data collected from these heterogeneous sources. It is clear that there needs to be more focus on open, distributed platforms that facilitate the integration of sensor data and understanding the exact meaning of integrated data, given the lack of research efforts on the application or platform level for ensuring integrated web-accessible data systematically. Ontologies enable communication among people or heterogeneous and widely spread application systems by creating a common understanding of the structure of information.

Semantic operable agriculture information system objectives include: (i) collecting domain-specific data about particular agricultural products through ontology-based data collection forms generated by domain stakeholders using crop specific ontologies; (ii) gathering and visualizing stream data concerning site-specific parameters of particular agricultural products through wireless sensor networks; (iii) using crop-specific ontologies to construct mapping rules that produce domain-specific linked data; (iv) using various databases and files, such as relational databases, graph databases, XML documents, RDF documents, etc.; (v) interoperability between distinct types of software applications thanks to the use of web services and APIs; and (vi) presenting structured, semantic, and well-defined data about a particular agricultural product using open standards in appropriate formats such as JSON, RDF/JSON, RDF/XML, N-triples, Notation 3, XML, HTML, CSV, and Excel. Due to the fact that all the objectives of the platform are derived from crop-specific ontologies, you might describe them as the platform's main pillars. We present a semantic interoperability framework adhering to all the objectives above for the agriculture domain.

Figure 9.8 shows four layers of the proposed approach, which provides solutions for interoperability issues in terms of publishing data in appropriate formats from the agricultural domain. These interrelated layers are (i) physical layer, (ii) data content layer, (iii) processing layer, and (iv) user interface, respectively.

Physical layer: The use of IoT devices in agriculture has been quite common because these devices can gather environmental and soil-specific data such as the pH of the soil, the salinity, the moisture of the soil, and the local climate (precipitation,

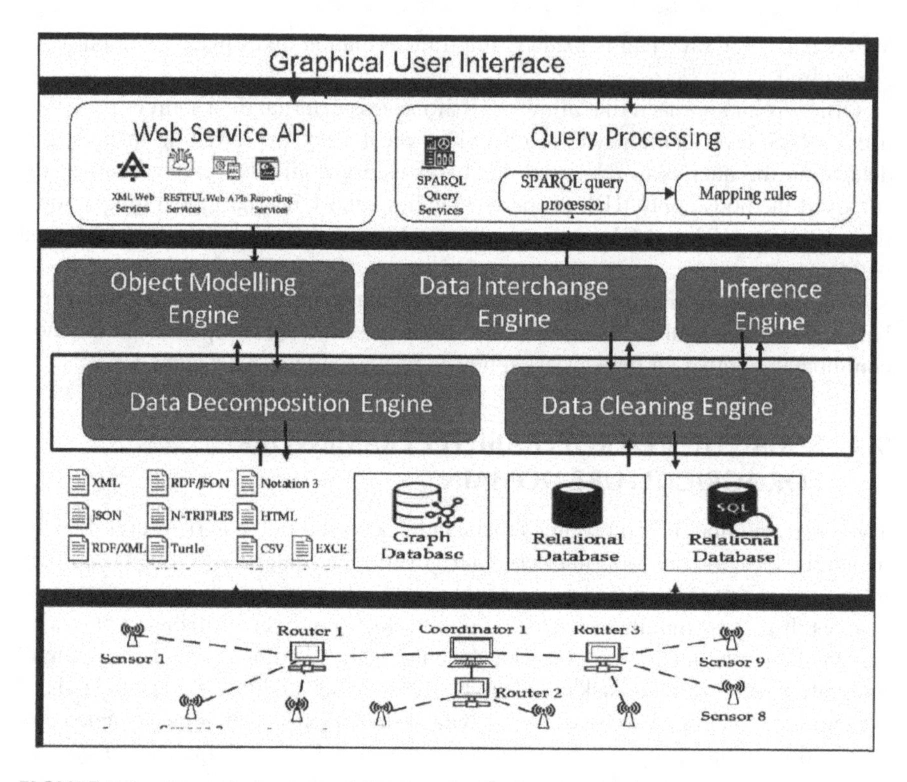

FIGURE 9.8 Semantic interoperability framework for agriculture domain.

temperature, wind speed, relative humidity). The physical layer is the core component of the proposed approach, and this consists of IoT devices arranged into wireless sensor networks (WSNs), which are controlled by a single controller. In agricultural applications, WSNs come in two varieties: terrestrial WSNs and underground WSNs.[24] There are several agricultural WSN standardizations that can be used for these applications, such as 6LoWPAN, WirelessHART, and ZigBee. The proposed approach does not restrict the choice of WSN types and standards.

Data content layer: This layer is concerned with decomposing raw sensor data streams, cleaning raw sensor data, modelling data, and storing data. Raw sensor data can be transformed into an appropriate format mapped to object models and stored in a variety of databases with the help of the physical layer. There are four sublayers in this layer: data storage engine (DSE), data decomposition engine (DDE), data cleaning engine (DCE), and object modelling engine (OME). The next step is to describe these four sub-layers in detail.

Data storage engine: Three different storage options are available for storing data obtained from sensors, including cloud database, relational database, and graph database.

Data decomposition engine: A sensor is a low-cost device used to measure site-specific and environmental data in agricultural applications. A sensor measures at least one value but can also measure other values. Due to the variety of sensors connected to router devices, the proposed approach recommends that sensors on WSNs be defined using key-value pairs. In DDE, raw sensor data is decomposed and sensor definitions, sensor measurement values, and network router definitions are extracted. Sensor measurement value is matched to its unit by using database or file data and mapped sensors to the measurement value for the creation of appropriate object models of sensors and routers that can be used by analysis tools.

Data cleaning engine: As a result of transmission problems, sensors frequently report missing values, and the problem is widely accepted within WSNs. Sensor measurements and anomalous values are handled by the data cleaning engine. Detecting anomalies and estimating missing sensor measurement values can be crucial when it comes to critical systems.

Object modelling engine: This layer is responsible for creating abstract object models of sensor data in order to map them to databases and ontologies.

Data interchange engine: As mentioned earlier, ontologies originate in computer science for exchanging and communication among various software agents. This component is responsible creating ontologies based on object model created in OME related crop and instantiate them with sensor values measured in physical layer. Domain experts are involved in this process. Multiple ontologies also need to map to create the mapping rule which further used by inference engine.

Agricultural ontology concepts are also linked to data obtained from IoT devices using mapping rules. Object models for sensor data types, data streams, and sensor object models are stored in RDF format and can be retrieved and manipulated using SPARQL. This allows the interchange of sensor data types, data streams, and sensor object models. Performing reasoning on semantic technologies can reveal the common knowledge of a domain by mapping database objects to high-level abstract models created within any programming language.

Inference engine: An inference engine applies logic rules to a knowledge base of ontologies to derive new information. A tool of choice to improve interoperability is inference. It can be used to discover new relationships, analyse data automatically, or manage knowledge within an information system. As well as discovering possible inconsistencies in the integrated data, inference-based techniques are important. The data set to be considered may contain the relationship (AtmosphericPressure isA WhetherParamter) and ontology declarations (WhetherParamter isA ClimateOfTheSite) as shown in Figure 9.9. However, the statement (AtmosphericPressure isA ClimateOfTheSite) was not originally part of the data set. It can also be said that the new relationship was discovered. Through inferencing, it is possible to discover the properties of concept "ClimateOfTheSite" properties "sensorId", "DataType" will also be attached to "AtmosphericPressure".

Query processing: An RDF triple containing variables is called a graph pattern, such as subject, object, and property. In many ways, SPARQL is similar to SQL because it is inspired by it. In SPARQL, there are three kinds of patterns disjunctions, conjunctions, and optional patterns. The subject, predicate, and object of a triple pattern can be variables, similar to triples in RDF. Figure 9.10 shows sensor data ontology with SPARQL representation for query.

Web service API: Web services, particularly RESTful web services, can assist software applications across platforms by enabling them to request, access and

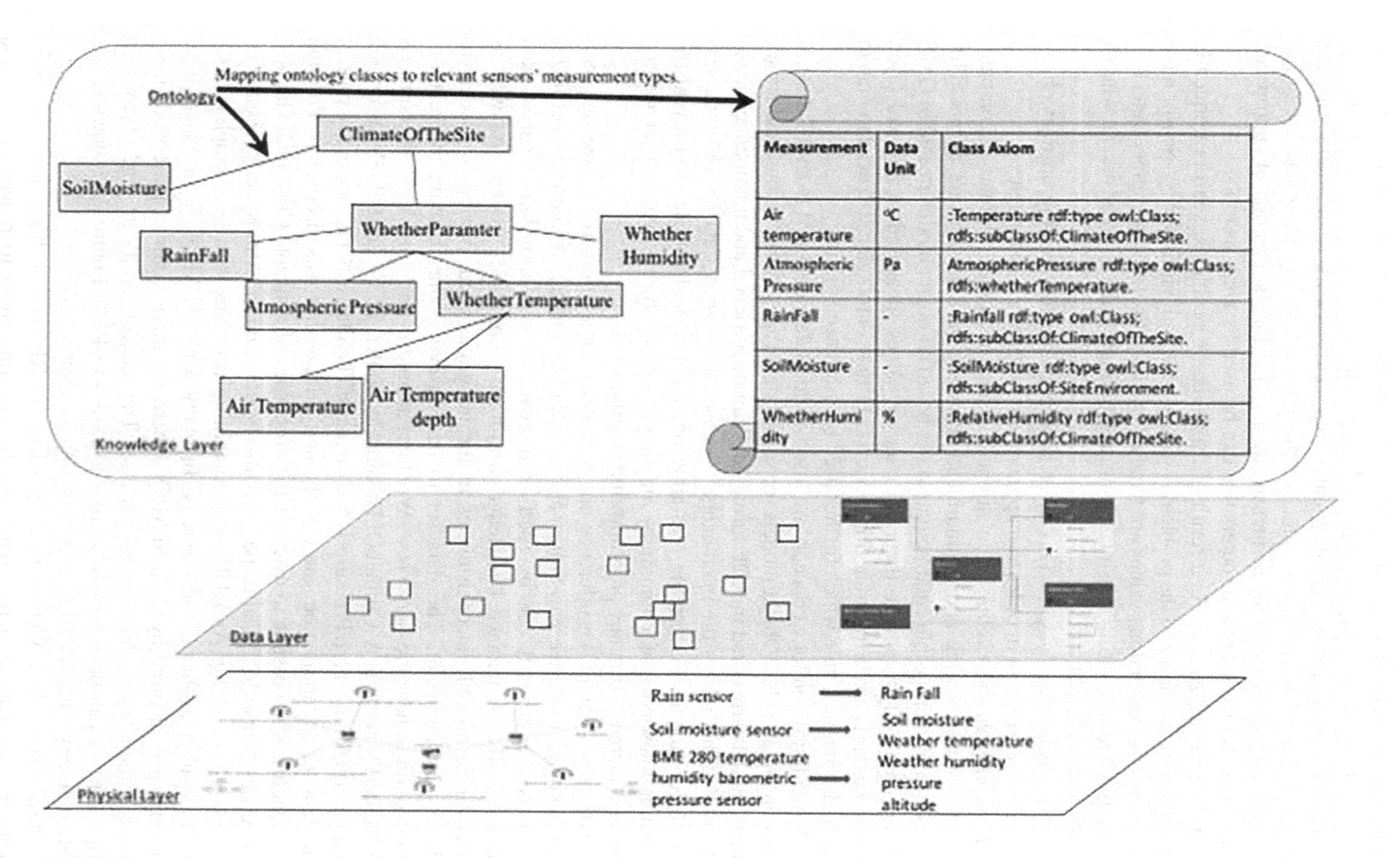

FIGURE 9.9 Layer-based method for semantic interoperability for agricultural domain.

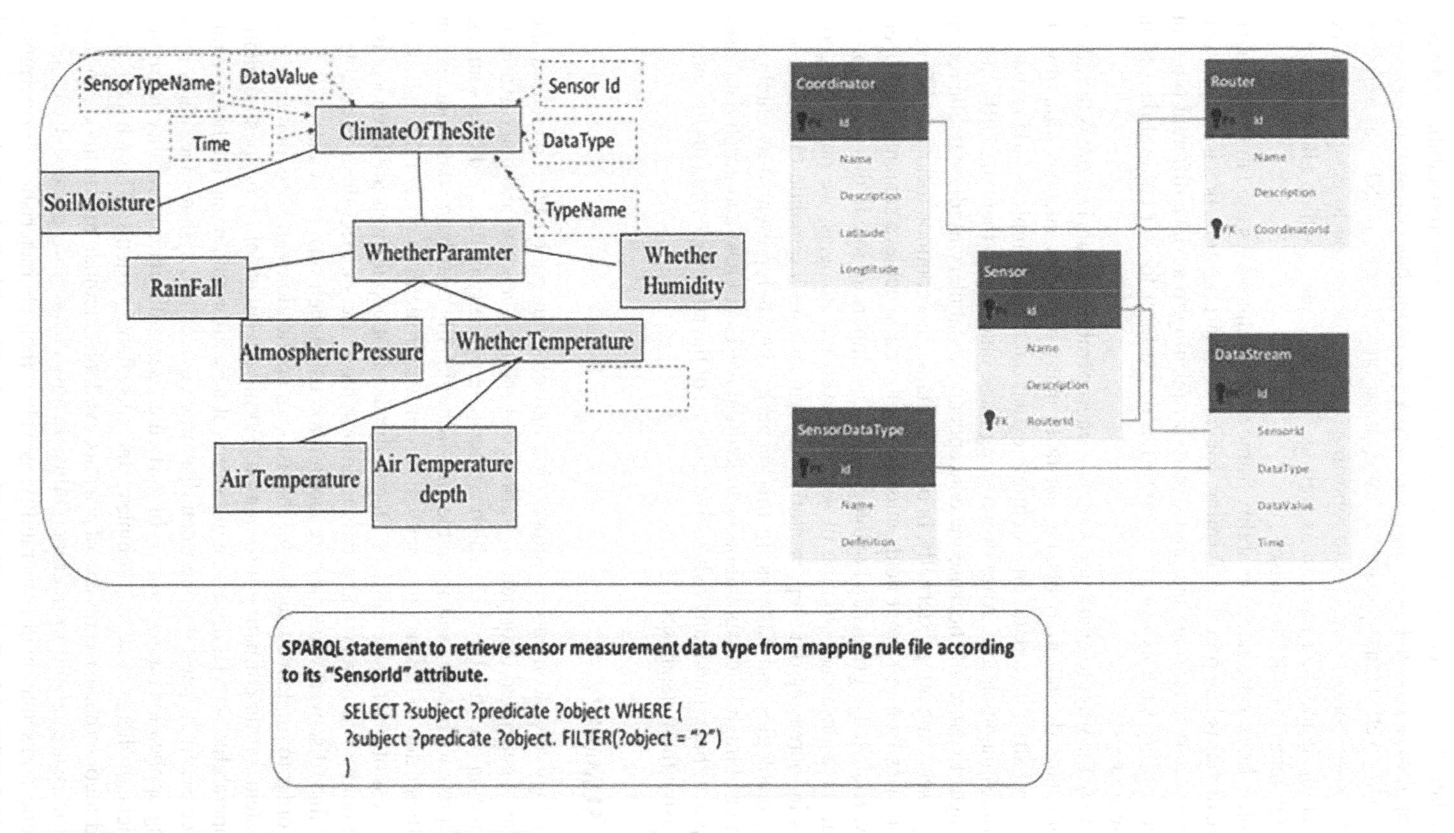

FIGURE 9.10 Sensor data ontology with SPARQL representation for query.

manipulate data, which can then be easily handled. The interoperability between these applications is not only an issue across platforms but also within the same platform as well. Consequently, many applications rely on REST web services, web APIs, or both. There is also reporting, which allows you to design and manage well-designed data reports, as well as SPARQL query, which allows running SPARQL queries on RDF datasets.

Table 9.1 describes interoperability frameworks/technologies used in various agriculture domains like precision farming, IoT-based smart farming, precision livestock farming, smart irrigation, and precision dairy farming to achieve different types of interoperability such as semantic, syntactic, or platform for agricultural data from diverse sources.

Among the ways to guarantee semantic interoperability in agriculture is the adoption of standards and protocols. Farmer faces many obstacles, such as initial investment, semi-structured data, unstructured textual data, and complex systems that are sometimes incompatible. After overcoming these challenges, there are still internal challenges to overcome because the concepts and vocabulary used across the region must preserve meaning externally. In order to maintain a common meaning, different vocabularies from different fields must be combined to achieve semantic interoperability. Knowledge graphs and ontologies play an important role in this scenario. As a result of widespread adoption of semantic web technologies, we observed good results in increasing data representation in formats with a semantic focus by combining ontologies and patterns. In this scenario, data is represented as ontologies, which encourages the exploration of linked databases (LODs) and graph databases (Virtuosos) but without any exhaustive analysis of the advantages and disadvantages of utilising these databases.

9.5 SUMMARY

Technology advancements in information and communication have increased the amount of generated information. Various stakeholders generate and make available information over the Internet, which creates heterogeneity of data and data sources. Attempting to discover and assemble data from globally distributed data sources can be difficult due to this heterogeneity. It has proved difficult to govern the discovery and access of heterogeneous data. Due to the fact that applications are intended as a single point of discovery and access to information, users expect them to be able to easily find (discover) what they are looking for, using their own language.

In order to implement such a mechanism, an application will need an interoperable system for disseminating information. Taking advantage of ontologies, semantic interoperability will facilitate the discovery of new information and facilitate seamless access. Interoperability between disparate devices and systems facilitates integrating applications across them. In particular, semantic interoperability is highly beneficial in heterogeneous, complex, and unscalable systems. This chapter presented a comprehensive review of various available methods and methods to deal with the issues related to interoperability. In this chapter, we discussed technical, syntactic, and semantic interoperability, as well as their characteristics. Furthermore, we demonstrated an ontology-based information system life cycle and demonstrated

TABLE 9.1
Interoperability in Agriculture Literature Review

Sr. No	Ref No	Technology Used/Tools	Data Sources	Interoperability Type	Summary
1	[26]	FIWARE Cloud, WSN, Widhoc GPRS Node	IoT Sensors	Syntactic	Application for smart irrigation using cloud service FIWARE having various advantages
2	[27]	OGC, XML, JSON, REST API	Wireless sensors, GPS sensors	Syntactic	Review of open standards and role of interoperability in precision livestock farming for better decision support system
3	[28]	Space4Agri, OGC standard interoperable SDI architecture, Geospatial Data	Remote Sensing, Unmanned Aerial Vehicle (UAV), Smart App	Platform	Development of framework and smart App using geospatial data and time series for precision farming
4	[29]	OntoAgroHidro, reuse of Cuahsi, CREON, SWEET	External organizations like governmental institutions and partners	Sematic	Interoperability issues, ontology to represent knowledge in agriculture field to study impact
5	[30]	Hive, MongoDB and Cassandra, agricultural big data	Sensors, user data, logs, data collected from queries on different interfaces	Syntactic	Agricultural data warehouse implementation and its evaluation using big data
6	[31]	ATLAS Registry, TLS	In field sensors, sensor platforms, nutrient and livestock behaviour data	-	Implementation of interoperability architecture for providing technical solutions by interconnecting components
7	[32]	OLAP, semantic warehouse	Sensors, agricultural information systems, dairy herd management systems, milk and feed quality data	Sematic	Making active semantic data warehouse using semantic technologies for precision dairy farming
8	[33]	Java, VRaptor, PostgreSQL	Sensors, satellite images, maps, harvest data, climate data, sample grid	Syntactic	Web interface for managing agricultural data and integrating applications for interoperability in precision agriculture
9	[34]	SOLAP, Hadoop	Satellite images, field sensors, weather data, crop registers, boundary data	Syntactic	Architecture for heterogeneous data integration in precision agriculture
10	[1]	FIWARE, IoT, s NGSI-LD Broker, JSON-LD, MEGA, MongoDB	Sensors, actuators, open information	Platform	Integrated data model for interoperability for smart irrigation
11	[35]	SPARQL, SQL, RDBMS, WSN, XML, REST	Hazelnut trait ontology, crop specific trait ontology, sensors	Syntactic	Usage of APIs, data integration approaches and web services for semantic and syntactic interoperability
12	[36]	ISOXML and JSON-LD, OWL, RDFS, SKOS	Sensors, FMIS, GIS, Satellite images, remote sensing, weather, logistic, harvest data	Syntactic	Providing conceptual interoperability framework for connecting various components used in farming

the use of semantic technologies. The use of ontologies for semantic interoperability was also described through a use case from the agriculture domain.

NOTES

1. RDF—Semantic Web Standards (w3.org)
2. OWL—Semantic Web Standards (w3.org)
3. SPARQL Query Language for RDF (w3.org)
4. https://cacm.acm.org/magazines/2021/2/250085-a-review-of-the-semantic-web-field/fulltext

REFERENCES

1. López-Morales, J. A., Martínez, J. A., & Skarmeta, A. F. (2020). Digital transformation of agriculture through the use of an interoperable platform. *Sensors*, 20(4), 1153.
2. Zhao, Y., Liu, Q., & Xu, W. (2017). Open industrial knowledge graph development for intelligent manufacturing service matchmaking. In *International conference on industrial informatics-computing technology, intelligent technology, industrial information integration (ICIICII)*. IEEE, pp. 194–198.
3. Tolk, A., & Muguira, J. A. (2003). The levels of conceptual interoperability model. In *Proceedings of the 2003 fall simulation interoperability workshop*, vol. 7. Citeseer, pp. 1–11.
4. Rezaei, R., Chiew, T. K., Lee, S. P., & Aliee, Z. S. (2014). Interoperability evaluation models: A systematic review. *Computers in Industry*, 65(1), 1–23.
5. Lutz, M., Sprado, J., Klien, E., Schubert, C., & Christ, I. (2009). Overcoming semantic heterogeneity in spatial data infrastructures. *Computers & Geosciences*, 35(4), 739–752.
6. Albouq, S. S., Sen, A. A. A., Almashf, N., Yamin, M., Alshanqiti A., & Bahbouh, N. M. (2022). A survey of interoperability challenges and solutions for dealing with them in IoT environment. *IEEE Access*, 10, 36416–36428. doi: 10.1109/ACCESS.2022.3162219
7. Garijo, D., & Osorio, M. (2020). OBA: An ontology-based framework for creating REST APIs for knowledge graphs. In *International semantic web conference*. Cham: Springer, pp. 48–64.
8. Zerger, A., & Smith, D. I. (2003). Impediments to using GIS for real-time disaster decision support. *Computers, Environment and Urban Systems*, 27(2), 123–141.
9. Sotoodeh, M., & Kruchten, P. (2008, April). An ontological approach to conceptual modeling of disaster management. In 2008 2nd Annual IEEE Systems Conference (pp. 1-4). IEEE.
10. Sánchez, D. M., Cavero, J. M., & Martínez, E. M. (2007). The road toward ontologies. In *Ontologies*. Boston, MA: Springer, pp. 3–20.
11. Gruber, T. R. (1993). A translation approach to portable ontology specifications. *Knowledge Acquisition*, 5(2), 199–220.
12. Grüninger, M. (2009). Using the PSL ontology. In the Handbook *on ontologies*. Berlin, Heidelberg: Springer, pp. 423–443.
13. López, M. F., Gómez-Pérez, A., Sierra, J. P., & Sierra, A. P. (1999). Building a chemical ontology using methontology and the ontology design environment. *IEEE Intelligent Systems and Their Applications*, 14(1), 37–46.
14. Brickly, D., & Guha, R. V. (2004, February). RDF Schema. *W3C Specification*.
15. Arvor, D., Durieux, L., Andrés, S., & Laporte, M. A. (2013). Advances in Geographic Object-Based Image Analysis with ontologies: A review of main contributions and limitations from a remote sensing perspective. *ISPRS Journal of Photogrammetry and Remote Sensing*, 82, 125–137.

16. Bharambe, U., Durbha, S. S., King, R. L., Younan, N. H., & Kurte, K. (2016). Use of geo-ontology matching to measure the degree of interoperability. In 2016 IEEE International Geoscience and Remote Sensing Symposium (IGARSS), China, pp. 7601–7604. doi: 10.1109/IGARSS.2016.7730982.
17. Jung, C. T., Sun, C. H., & Yuan, M. (2013). An ontology-enabled framework for a geospatial problem-solving environment. *Computers, Environment and Urban Systems*, 38, 45–57.
18. Visser, U. (2004). General approach of buster. In *Intelligent information integration for the semantic web*. Berlin, Heidelberg: Springer, pp. 37–51.
19. Wache, H., Voegele, T., Visser, U., Stuckenschmidt, H., Schuster, G., Neumann, H., & Hübner, S. (2001). Ontology-based integration of information-a survey of existing approaches. In *Ois@ ijcai*.
20. Hitzler, P. (2021). A review of the semantic web field. *Communications of the ACM*, 64(2), 76–83.
21. Ekaputra, F., Sabou, M., Serral Asensio, E., Kiesling, E., & Biffl, S. (2017). Ontology-based data integration in multi-disciplinary engineering environments: A review. *Open Journal of Information Systems*, 4(1), 1–26.
22. J. Angele and M. Gesmann, "Data Integration using Semantic Technology: A use case,"2006 Second International Conference on Rules and Rule Markup Languages for the Semantic Web (RuleML'06), Athens, GA, USA, 2006, pp. 58-66, doi: 10.1109/RULEML.2006.9.
23. Lee, J. Y., & Kim, K. (2007). A distributed product development architecture for engineering collaborations across ubiquitous virtual enterprises. *The International Journal of Advanced Manufacturing Technology*, 33(1), 59–70.
24. Imran, M., & Young, R. I. (2016). Reference ontologies for interoperability across multiple assembly systems. *International Journal of Production Research*, 54(18), 5381–5403.
25. Ojha, T., Misra, S., & Raghuwanshi, N. S. (2015). Wireless sensor networks for agriculture: The state-of-the-art in practice and future challenges. *Computers and Electronics in Agriculture*, 118, 66–84.
26. López-Riquelme, J. A., Pavón-Pulido, N., Navarro-Hellín, H., Soto-Valles, F., & Torres-Sánchez, R. (2017). A software architecture based on FIWARE cloud for precision agriculture. *Agricultural Water Management*, 183, 123–135.
27. Bahlo, C., Dahlhaus, P., Thompson, H., & Trotter, M. (2019). The role of interoperable data standards in precision livestock farming in extensive livestock systems: A review. *Computers and Electronics in Agriculture*, 156, 459–466.
28. Bordogna, G., Kliment, T., Frigerio, L., Brivio, P. A., Crema, A., Stroppiana, D., ... Sterlacchini, S. (2016). A spatial data infrastructure integrating multisource heterogeneous geospatial data and time series: A study case in agriculture. *ISPRS International Journal of Geo-Information*, 5(5), 73.
29. Bonacin, R., Nabuco, O. F., & Junior, I. P. (2016). Ontology models of the impacts of agriculture and climate changes on water resources: Scenarios on interoperability and information recovery. *Future Generation Computer Systems*, 54, 423–434.
30. Ngo, V. M., Le-Khac, N. A., & Kechadi, M. (2019, June). Designing and implementing data warehouse for agricultural big data. In *International conference on big data*. Cham: Springer, pp. 1–17.
31. ATLAS, Deliverable D3.2 service architecture specification. (2020). www.atlas-h2020.eu/wp-content/uploads/2020/06/ATLAS-D3.2-Service-Architecture-Specification.pdf.
32. Schuetz, C. G., Schausberger, S., & Schrefl, M. (2018). Building an active semantic data warehouse for precision dairy farming. *Journal of Organizational Computing and Electronic Commerce*, 28(2), 122–141.

33. Bazzi, C. L., Jasse, E. P., Magalhães, P. S. G., Michelon, G. K., de Souza, E. G., Schenatto, K., & Sobjak, R. (2019). AgDataBox API–integration of data and software in precision agriculture. *SoftwareX*, 10, 100327.
34. Gallinucci, E., Golfarelli, M., & Rizzi, S. (2019, August). A hybrid architecture for tactical and strategic precision agriculture. In *International conference on big data analytics and knowledge discovery*. Cham: Springer, pp. 13–23.
35. Aydin, S., & Aydin, M. N. (2020). Ontology-based data acquisition model development for agricultural open data platforms and implementation of OWL2MVC tool. *Computers and Electronics in Agriculture*, 175, 105589.
36. Bökle, S., Paraforos, D. S., Reiser, D., & Griepentrog, H. W. (2022). Conceptual framework of a decentral digital farming system for resilient and safe data management. *Smart Agricultural Technology*, 2, 100039.
37. Sure, Y., Staab, S., & Studer, R. (2009). Ontology engineering methodology. In Handbook on ontologies (pp. 135-152). Berlin, Heidelberg: Springer Berlin Heidelberg.

10 Design and Implementation of a Short Circuit Detection System Using Data Stream and Semantic Web Techniques

Ing. Luis Ernesto Hurtado González
and C. Amed Abel Leiva Mederos

CONTENTS

10.1 INTRODUCTION

Electrical power systems have become one of the fundamental bases of human development; they are in charge of bringing electrical energy from generating plants to large consumption centers such as cities and industries. Due to this, electrical systems are constantly changing, which allows us to provide a reliable and safe service. In this process, there are failures that are caused by natural phenomena, such as storms, lightning, deterioration of insulation, and even birds, trees, and human accidents. An electrical system presents faults of the series type (breakage of conductors)

DOI: 10.1201/9781003310792-10

and of the parallel type (short circuit to ground or between phases). The fault that causes the most damage in a power system is the short circuit, which occurs when energized conductors of different phases come into contact with each other or with ground. Finding short circuits on transmission lines and poles has been a matter of concern for network operators, as lines are known to be characterized by immense lengths and large numbers of poles. In case of failure, it is difficult to know the exact point where it is generated. This causes great uncertainties and inconveniences because it is absolutely necessary to review section by section to locate the point that caused the anomaly.

The development of science and technology in various fields has allowed the use of virtual tools for the management and control of electrical systems. These are mainly implemented in developed countries, which is an important advantage for the rapid detection and location of faults in electrical systems and their solution. The Internet of Things (IoT) constitutes the technological basis for the use of these tools. Innovation and development in leading technology companies lead to a marked difference between their use in the most developed countries and those that do not have the necessary advanced digital infrastructures, as is the case in Latin America. In Cuba there are many failures in the national electrical system, mainly caused by short circuits. There are no efficient methods to monitor and locate them. When failures occur in the national electricity system, the response time of the electric company is fast, which affects a large number of users. For this reason, it is necessary to find effective methods to address this problem.

10.2 RELATED WORK

It is necessary to relate different concepts linked to time, values, stream, location, devices, and quantities to represent the flows of information and data of IoT. There are models for the representation of sensors and their observations, which appear in the description of the device. The Semantic Sensor Network (SSN) ontology is the most typical model and describes the sensors with their properties, stimuli, displays, systems, and observations.[1] On the other hand, the Sensor, Observation, Sample, and Actuator (SOSA) ontology is a lightweight kernel for SSN that provides concepts for sensors with peculiar characteristics and observation values.[2] There is another IoT model, the IoT-Lite, which consists of a lightweight model that aims to speed up semantic annotation, query time, and data processing. This model arises from the Internet of Things – Architecture (IoT-A).[3] The approach defined from the analysis of SSN, IoT-Litem and SOSA is appropriate for sensor discovery and has more representativeness in detection devices because they lack specific concepts for stream aggregation and annotation.

There are certain location models like Geo that make it easy to find[1] IoT devices. Geo is a very popular model known for representing location data in Resource Description Format (RDF) format. Some essential and basic terms the ontology offers for use in RDF for the description of measurements are altitudes, latitudes, and longitudes. Using RDF as a carrier for altitude, latitude, and longitude facilitates the ability to mix and match data across domains, as well as the description of features found on the map (for example, geospatial queries for sensors, platforms, systems, and/or data or implementations). Among the variants used for querying information and representing geospatial data is GeoSPARQL, which constitutes a data pattern linked to the semantic web of the Open Geospatial Consortium (OGC).[4]

The ontology of time,[2] as its name indicates, is used to represent the units linked to the time factor and constitutes one of the most studied, used, and well-known semantic models. Due to the typicity that characterizes it, it has a defined vocabulary to classify and represent information on topological relationships (ordering), duration, and period (date and time).

Data linked to time can be expressed in different ways, such as by means of a conventional clock, geological time, Unix time, and other systems that can be used as reference standards. Time ontologies have been used for the annotation, query, study, and use of stream data and, similarly, to evaluate other units, dimensions, quantities and values. In this sense, the Quantity Units (QU) ontology[3] is one of the most studied and recognized ontologies in this field of research. It has a development that allows the admission of several different users of the systems modeling language (SysML).

As in many other aspects, quality is one of the most important elements. In the present investigation, the quality of information (QoI) constitutes an outstanding element of the annotation of the data stream, since the use of incorrect data can lead to important expenses as a direct consequence.[5] If it is about the quality of the information, the metrics are significant elements to detail the details. There are five common metrics: consistency, completeness, correctness, timeliness, plausibility,[6] and security.[7]

The previously mentioned ontologies constitute an important element of support in the annotation of data stream, but as a deficiency, it should be noted that they do not fully address the necessary and essential concepts for the correct analysis of the data. The incorporation of stream that allow the use of information processing time when consulting data is considered opportune. There are very few ontologies that represent data in stream. Streaming annotation ontology (SAO) is one of the representatives incorporating IoT data stream, based on previous studies: TimeLine,[8] PROV-O,[9] Social Security Number,[1] and event ontology.[8] The related concepts through StreamEvent, StreamData, Observation, StreamAnalysis, Segment, and Sensor make it possible to accurately describe the temporal concepts of this ontology. Among its benefits, with the StreamData class, SAO provides a data stream as a time point or segment and describes the result of the observation as an event with the StreamEvent class.[10]

10.3 PROPOSED WORK

10.3.1 System Design

The main pattern of information is directed toward the modeling of the observations of the streams, their analysis, and the events that are discovered as a result of it, which are classified into four classes. They show the concepts of StreamObservation, IoTStream, Event, and Analytics. As shown in Figure 10.1, the main class that the rest of the classes bind to is IoTStream. This abstraction evidences streams that are

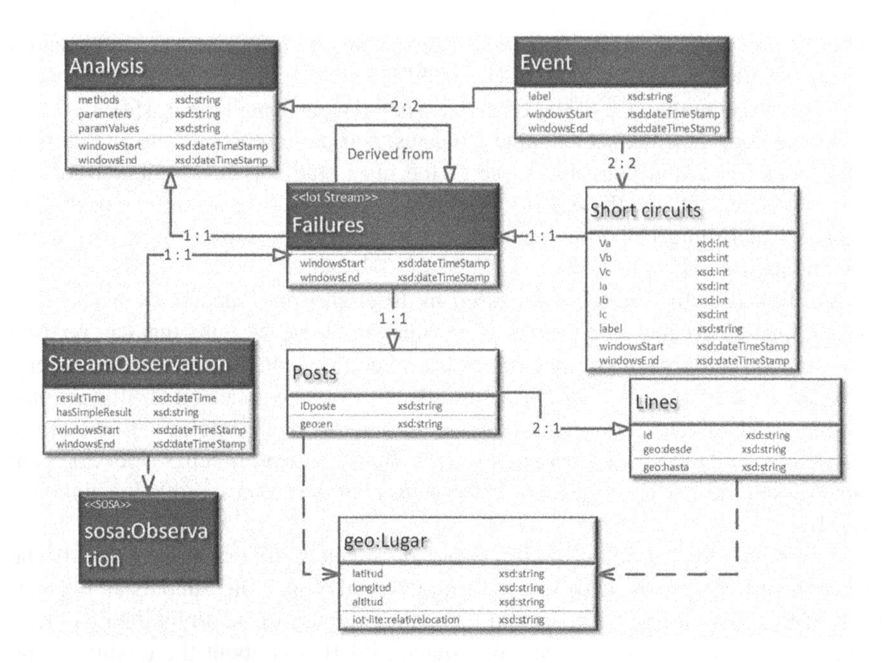

FIGURE 10.1 IoTStream classes and properties.

born from an IoT data source. It has annotation properties that capture the useful life of the streams that would be used as a reference instead of the actual consumption of an application. The properties of this annotation are streamStart and streamEnd. The vision of an IoTStream such as water currents, can branch off into different currents and, therefore, derive from another IoTStream.

The value of watching a broadcast can take different forms: first, atomic, which is represented by a data point; and second, massive figuration, which is represented by a vector of data points. StreamObservation includes an instant timestamp that keeps track of when the observation was captured. Studies reveal that if popular ontologies are taken into account, SOSA provides a class that meets the requirements for obtaining sensor information, which is the sosa:Observation class and its data type properties, sosa:hasSimpleResult and soda:resultTime. Similarly, it is important to consider the temporal aspect of an observation as an interval or window, sosa:Observation provides the property to classify according to sosa:phenomenTime data type for this purpose and is linked to the time:TemporalEntity class of the popular time ontology. The observation class has been extended with a subclass, StreamObservation, to contain properties of direct data types that represent time windows. These data are obtained in the windowStart and windowEnd data properties, which show the start and end, respectively, of the window. Although the use of StreamObservations is important, it is also important to segregate it from the rest of the metadata because when establishing comparative analyzes it is concluded that the number of established instances would be significantly higher compared to the number of IoTStreams, and for this reason, the sosa:Observation's instantiation scope is provokingly and intentionally kept to a minimum.

StreamObservation that belong to IoTStreams can be the result of sensor readings or the result of an analysis process. If IoTStream is studied through a data analysis process, the Analytics class accurately recreates the techniques and methods used, whether it is one of its processes or several of them, so that it appears as a string of vectors with the properties of ParamValues data, methods, and parameters. Naturally when analyzing the streams, the techniques, methods, procedures, and algorithms used are captured by data property methods. The set of properties that revolve around the data and parameters for these methods is captured as a string of vectors, which is why the first element in the method vector corresponds to the first element in the parameter vector. It should be noted that many parameters are established by previously defined methods, so the element that corresponds to the data property of said parameter can also be an array of parameters. The values that are given to each parameter are captured in the ParamValues data property. There is the possibility of keeping the analysis process used on an IoTStream active through a temporary window during the lifetime of an IoTStream; that is why the windowStart and windowEnd data properties are used in similar situations.

The generation of events through the use of data analysis is one of the elements addressed. For this purpose, the Analytics class is used, which defines the process to generate events that manifest an IoTStream through detectedFrom, which is used in cases of classification or grouping. Among the characteristic properties of the event are those of generating labels for the description of the same and the time of completion.

10.3.2 Metrics and Ontology Documentation

The ontology is based on the vocabularies described in Table 10.1. The main classes contain the electrical network, the faults, and the line types; in addition the Sensor and ObservableProperty classes are presented with the energy variables and the elements that allow the detection of faults and the geographic coordinates. The properties of the object in the ontology are joined in the following way: classes/individuals. They are the most important elements of the ontology and act to manage short circuits in the network. The main properties of the object are listed in the proposed ontology in the table.

Rules and conditions:

- Messages (? P) ^ core: Goes (? P, 0) ^ Vb (? P, 0) ^ Vc (? P, 0) ^ core: Ia (? P, 0) ^ Ic (? wh? Ic2) ^ Ib (? P,? Ib) ^ swrlb: less than or equal to (? Ic2, 0) ^ swrlb: greater than or equal to (? Ib, 0) -> BiPhaseFailure (? P)
- (Ib some xsd: int [> "1" ^^ xsd: int]) and (Ic some xsd: int [> "1" ^^ xsd: int]) and (Ia some xsd: int [> "1" ^ ^ xsd:int]) and (Vb value 0) and (Vc value 0) and (Va value 0)

The first rule is that when a short-circuit is of the biphasic type, it meets certain conditions and if all of them are fulfilled, we are in the presence of a biphasic short-circuit.

TABLE 10.1
Properties of the Object, Domain, Range, and Description of the Main Property of the Object

Object Property	Domain	Range	Description
soda:doneObservation	bland: acting red: operations	soda:Sensor	It allows the relationship between the short-circuit control operations and the Sensor.
soda:resultTime	bland: acting soda: Remark	time time	It relates the operations and the observation of the short circuits during a certain time through the observed variables.
soda: observe	soda:ObservableProperty bland: acting Red lines red: data	soda:Sensor bland: acting	The sensor observes the observable properties (variables) in the lines. These variables are data observed in real time or through historical data. Action or decision making depends on these variables.
soda: owned observed	soda: Remark	soda:ObservableProperty	Relate observations to observable properties (variables).
soda:doneObservation	Sensor	StreamObservation	Link sensors to stream observations.
soda:hasFeatureOfInterest	soda: Remark bland: acting	soda:FeatureOfInterest	Observation and action have specific objects in an electrical system, these are lines, poles, transformers, generators, loads, transformers, etc.
soda:hosts	soda:Platform	soda:Sensor	All sensors are hosted on a free software platform called Sofia.
soda:madeByActuator	bland: acting red: data red: operations red: electric network red: faults	soda: actuator red: positions red: Protections red: operations bland: acting	Actuation, data, operations, mains, and faults are related to the actuator, which is related to positions, protections, operations, and actuation.
soda:madeBySensor	soda: Remark	soda:ObservableProperty network:Electric Network	Relationship between observations, observable properties and the entire electrical network.
soda: it was originated by	ssn: Stimulus	soda: Remark	Each short circuit is related to an observation.
Iot-stream:derivedFrom	IotStream	IotStream	Relations between two streams where one is generated by the other.
Iot-stream:detectedFrom	Event	IotStream	Events that affect the electrical network are detected in stream.
Iot-stream: analyzed by	IotStream	analytics	Each stream has an analysis service and an Artificial Intelligence method that manages it.
Iot-stream: provided by	IotStream	Iot-lite:Service	Each stream is included in a service that has its SPARQL interface and endpoint.
qoi: it has quality	IotStream	qoi:frequency	Determines how often the system is updated.
geo: in	IotStream	Place and fault	Bind the place and fail.

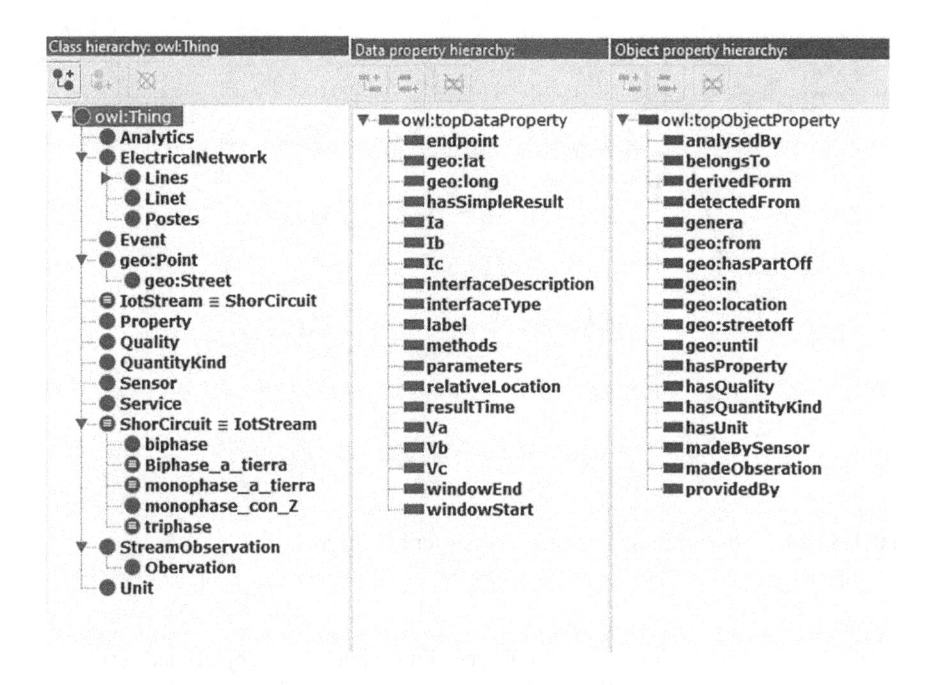

FIGURE 10.2 Classes, data property, and object property of the ontology.

The next point is a condition that belongs to a three-phase short circuit since it says that if all the voltages are at 0 and the currents are not 0.

The global model for IoTStream has 25 classes, 19 data properties, 18 object properties, 316 axioms, 228 logical axioms, and 88 declared axioms, 13 subclasses, 26 individual, 4 equivalent classes, inverseObjectProperties, 22 ObjectPropertyDomain, 36 DataPropertyDomain, 57 ClassAssertion, 27 ObjectPropertyAssertion, and 31 DataPrepertyAssertion. The classes, data property, and object property are shown in Figure 10.2.

10.3.3 System Architecture

For a system to accept IoTStream adoption, the notable system entities required, as shown in Figure 10.3, would be:

- Registry: It is in charge of storing the information related to IoTStream (triple storage) and to handle the queries, the SPARQL endpoint is used. IoTStream is used to store StreamObservations.
- Producer: Handles the registration of IoTStreams and the publication of the StreamObservations obtained from its sensors.
- Consumer: A service or application that manifests IoTStreams and performs StreamObservations through the chosen IoT service. The consumer can get the learned data for business understanding or preprocessing from StreamObservations, via filtering or aggregation.

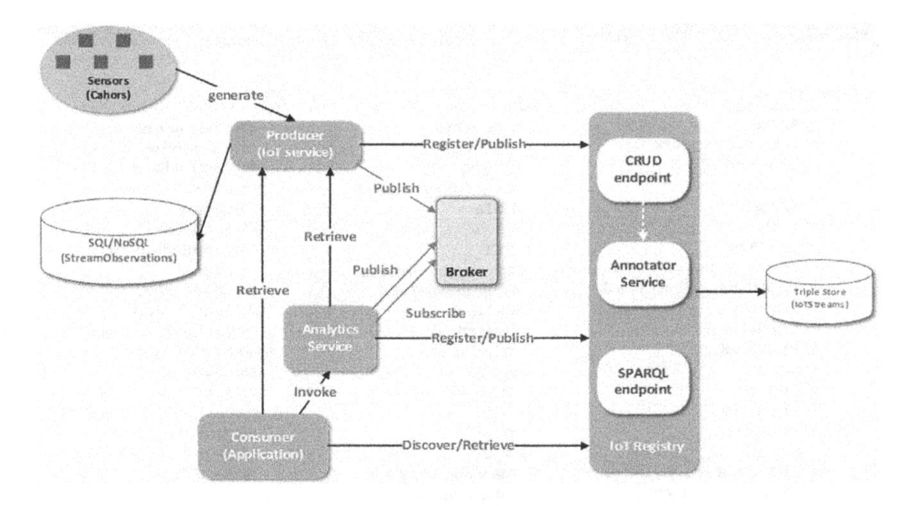

FIGURE 10.3 Interactions and system entities for IoTStream.

- Broker: An election to StreamObservations where data subjects and services can access real-time StreamObservations transmitted by a producer to the stream data broker. In this case, persistence depends on who consumes.
- Analysis Service: It is used by a consumer to absorb or create studied IoT-Streams, through key techniques for data analysis with a certain method or a set of these methods.

10.3.4 Validation

The Python libraries that were used to carry out this process of analysis, classification, clustering, and sampling are numpy (scientific calculations), pandas (data analysis), matplotlib (high-quality plot printing), and sklearn (K-means function).

The data to analyze is loaded. In this case, being a proof of concept, only a small amount of data is used. When a stream occurs it generates changing values like voltages and currents, but there are also other columns in the table like id, WindowsStart and WindowsEnd, geolocation which are necessary and very important data; however, at this moment, only Va values were needed: Vb, Vc, Ia, Ib, and Ic (to make the algorithm). A new variable containing only these values was created and later normalized.

With these values in the new variable already normalized, a method called "Jambú Code" was applied, which consists of creating different amounts of conglomerates and calculating how similar the individuals are within them and little by little capturing this information in a graph to decide how many clusters are necessary to solve the problem. In Figure 10.4 the result of applying this method to the selected data is presented.

The more similar the individuals are, the more distant the groupings that are formed will be, which is what is sought in grouping: the formation of well-defined

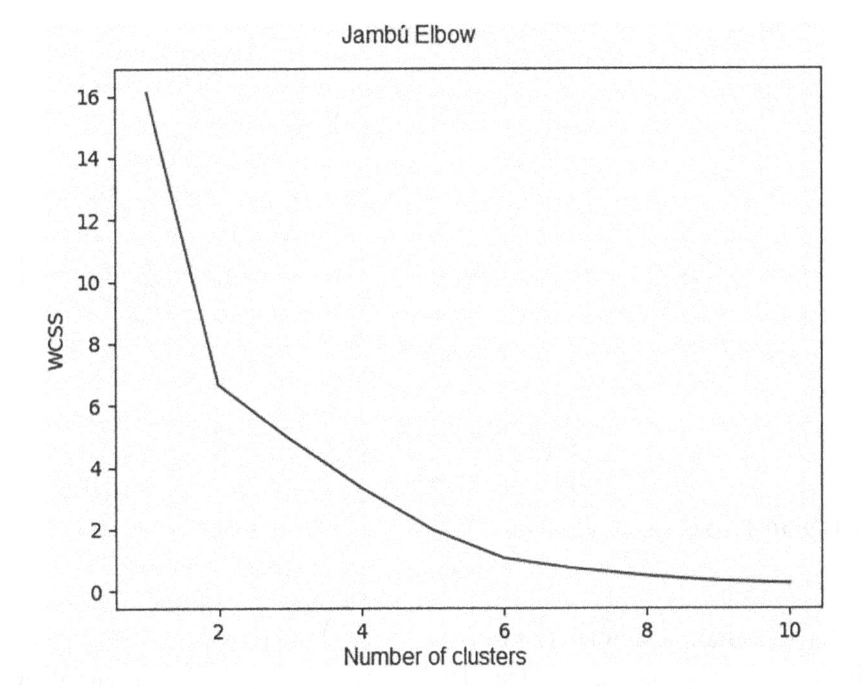

FIGURE 10.4 Jambú Elbow method.

groups that contain individuals whose distance between them is as small as possible. To measure this distance, the sum of the squares within each group (WCSS) is used. The value of WCSS decreases as the number of clusters increases, so when reaching 10 clusters the value of WCSS is already very small compared to what was had with one cluster, which is fine. But you'd also be doing a lot of partitioning, so you're looking for a point where the value of WCSS stops dropping drastically. In this case, that point is 2, so this is the optimal number of clusters to use.

Once the number of clusters into which the data is to be clustered is known, the K-means method is applied, and each stream the cluster belongs to is saved to a new column in the comma-separated values (CSV) file. Figure 10.5 continues with the principal component analysis to give an idea of how the clusters were formed.

Each point represents a stream and the colors to each group to which it belongs. The two clusters are well defined because they are comparing voltage and current values of the order of 110 volts and 20 amps, respectively, with respect to the other cluster, which are values at 0. As can be seen, if the clusters were made correctly, there are no mixed blue and green dots; the blue ones represent the measurement values that are in the proper operating ranges and the green ones represent those that have some zero value; this indicates that the probability of a failure occurring in the green clusters is very high. This algorithm is not to decide what kind of failure occurred or to say if there is a failure, although if it detects it, it is only to group the data and then gain precision. The Incremental Reasoning that ontology uses and the rules decide what happened.

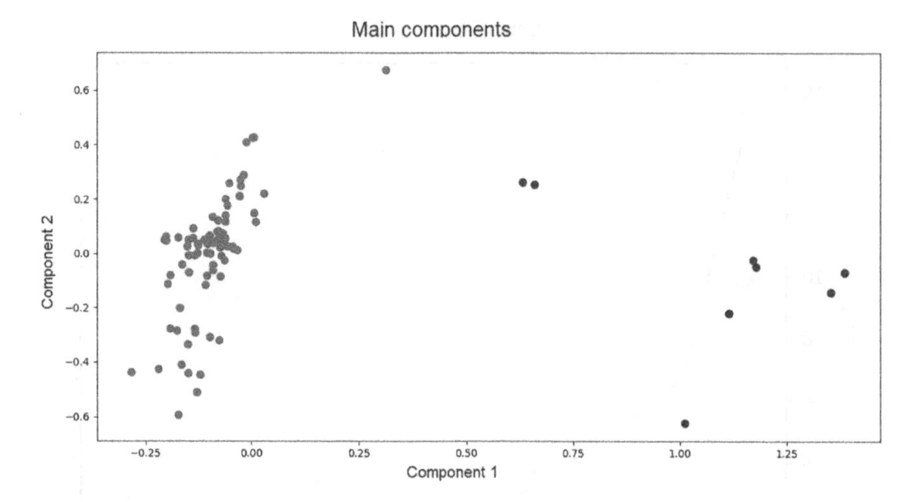

FIGURE 10.5 Data already grouped.

10.3.4.1 Database and Triplestore

The chosen NoSQL database is MongoDB because it is very easy to use and allows you to work with big data. All of the tools used do not save historical data, so if you want to do an analysis or apply an algorithm, MongoDB saves the data.

The file containing the data is loaded into the Virtuoso software (Figure 10.6), as is the ontology designed in Protégé (Figure 10.7). It is necessary to relate these data to their respective fields in the imported chart. For this to be possible, it is necessary to use RDF Mapping Language (R2RML) scripts.

R2RML is a W3C recommendation that allows you to specify rules for transforming relational databases to RDF. This RDF data can be materialized and stored in an RDF triplet management system (commonly known as a triplestore), where SPARQL queries can be evaluated. However, there are cases where materialization is not appropriate or possible, for example, when the database is frequently updated. In these cases, it is best to think of the data in RDF as virtual data, so that the SPARQL queries mentioned earlier are translated into SQL queries that can be evaluated in the original relational database management systems (DBMS).

The Virtuoso controller produces a file in the R2RML4 script mapping language. These automatically generated scripts create a generic SQL relational table ontology (Figure 10.8).

Virtuoso users have the ability to customize these basic scripts to map automatically generated column types to class properties from an external ontology or to map multiple table columns to a single RDF relation, among other things.

The R2RML graph generated by default is modified, with which the CSV columns where the instances (data) are located are matched with their respective predicate within the graph. In Figure 10.9 a query is displayed in the Virtuoso software where it is shown that the previous process is correct. It can be clearly seen that all the previous values of Ia, Ib, and Ic of the CSV with the data were correctly integrated in

CSV Import

Select parse options

Mode: Strict
Delimiter: ;
Quote: Double
Encoding: Default
Header Row: 1 zero if no header
Import from line: 2

Sample Data

#	_id	StreamStart	StreamEnd	Sensor	Va	Vb	Vc	Ia	Ib	Ic	Zf	lat	long	alt	frequency	idl	idp	Point	KMeans_Clusters
2	612c050b23b23c077c288613	2021-08-25T01:25:42:124Z	2021-08-25T01:30:42:124Z	2257	115	0	0	0	12	13	3	22	-79	220	0	35	36	Zulueta	0
3	612c050b23b23c077c288614	2021-08-25T01:30:42:124Z	2021-08-25T01:35:42:124Z	2254	116	0	0	0	21	25	3	22	-79	222	0	35	37	Zulueta	0
4	12c050b23b23c077c288615	2021-08-25T01:35:42:124Z	2021-08-25T01:40:42:124Z	2265	109	0	115	0	12	0	3	22	-79	221	0	35	38	Zulueta	0
5	612c050b23b23c077c288616	2021-08-25T01:40:42:124Z	2021-08-25T01:45:42:124Z	2248	110	110	110	23	12	12	3	22	-79	220	0	35	39	Zulueta	1
6	612c050b23b23c077c288617	2021-08-25T01:45:42:124Z	2021-08-25T01:50:42:124Z	2241	114	106	122	21	23	23	3	22	-79	225	0	35	40	Zulueta	1
7	612c050b23b23c077c288618	2021-08-25T01:50:42:124Z	2021-08-25T01:55:42:124Z	2257	0	0	0	23	12	13	3	22	-79	220	0	35	36	Zulueta	0
8	612c050b23b23c077c288619	2021-08-25T01:55:42:124Z	2021-08-25T02:00:42:124Z	2299	0	111	119	12	0	0	3	22	-79	221	0	35	42	Zulueta	1
9	612c050b23b23c077c28861a	2021-08-25T02:00:42:124Z	2021-08-25T02:05:42:124Z	2257	114	113	120	25	13	16	3	22	-79	220	0	35	36	Zulueta	1
10	612c050b23b23c077c28861b	2021-08-25T02:05:42:124Z	2021-08-25T02:10:42:124Z	2247	117	118	119	26	28	25	3	22	-79	222	0	35	44	Zulueta	1
11	612c050b23b23c077c28861c	2021-08-25T02:10:42:124Z	2021-08-25T02:15:42:124Z	2259	112	110	110	16	26	35	3	22	-79	220	0	35	45	Zulueta	1

Back Refresh Next

FIGURE 10.6 Loading the CSV into Virtuoso.

Virtuoso CONDUCTOR

logged in as dba | Log out | Home

Interactive SQL (ISQL)
WebDAV Browser
OpenLink Data Spaces
Virtuoso Start Menu
Documentation (web)
Tutorials (web)
Virtuoso Web Site
OpenLink Software
Version: 07.20.3233
Build: Jun 24 2021

Home | System Admin | Database | Replication | Web Application Server | XML | Web Services | Linked Data | NNTP

SPARQL | Sponger | Statistics | Graphs | Schemas | Namespaces | Views | R2RML | Quad Store Upload

Quad Store Upload

File* Browse... No file selected.
Resource URL*
Create graph explicitly
Named Graph IRI* http://localhost:8890/DAV

Cancel Upload

Copyright © 1998-2021 OpenLink Software

FIGURE 10.7 Importing the chart into Virtuoso.

RDB2RDF Triggers Status

Status	Message
00000	OK

SQL Relations (Tables) to RDF Statements (Predicate / Property Graph) Mappings

http://localhost:8890/schemas/CSV/qm-observation_csv
http://localhost:8890/schemas/CSV/qm-VoidStatistics

Sample Graph IRIs & Linked Data Entity URIs

RDF Document (Named Graph) IRIs:

Transient Views: http://localhost:8890/CSV#
http://localhost:8890/CSV/observation_csv/ID/1#this
http://localhost:8890/CSV/stat#
http://localhost:8890/CSV/stat#Stat
Materialized Views: urn:localhost:8890:CSV
http://localhost:8890/CSV/observation_csv/ID/1#this
Metadata Data Document (VoiD) URI/URL: http://localhost:8890/CSV/stat#
Linked Data Ontology URI: http://localhost:8890/schemas/CSV/

Start Over Back

FIGURE 10.8 Result of mapping the tables in RDF.

the graph as current data, likewise the values of Va, Vb, and Vc are voltages and Zf as impedance. Note that the SPARQL query describes all the observations that are made on the CSV.

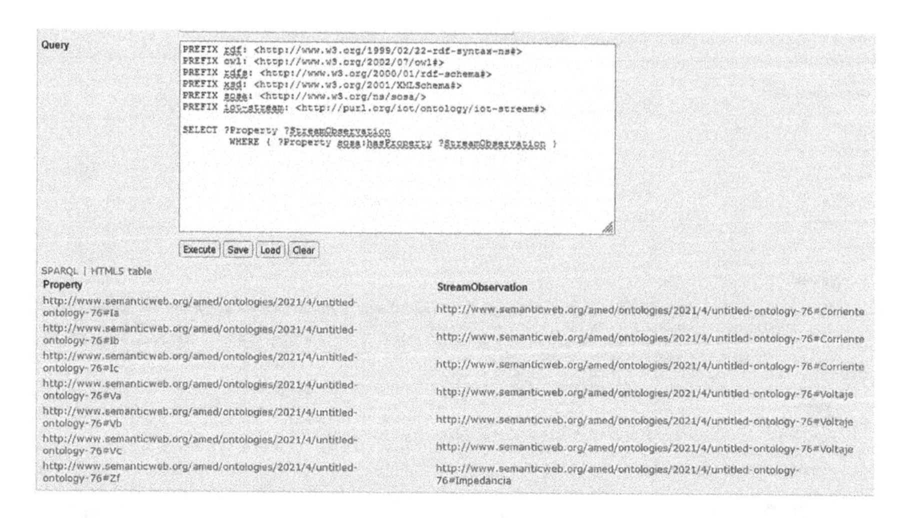

FIGURE 10.9 Query in SPARQL to demonstrate the integration of the data with the ontology in the triplestore.

10.3.4.2 SPARQL Queries

The Virtuoso tool has the ability to query data. Once the entire process described has been carried out, Virtuoso performs a reasoning, which is described in the ontology through rules that were already declared previously. These rules, based on conditions written in SWRL language, are able to analyze the conditions that imply a short circuit. The graph captures the instances that are in the CSV through OBDA and notes the items associated with the failure. In Figure 10.10 an example of a query is shown that asks the analysis system to return all the short circuits that occur, indicating that there is some type of short circuit in Zulueta. In this case, the geographical point where there is at least one short circuit is shown.

This is not enough because you need to know how many streams turned out to be short and your geographic location. To do this, a new query is made, modifying the classes to be displayed. In Figure 10.11 it can be seen that as a result of the inferences, there are only three currents that turned out to be faults.

It is important to know which sensor is reading the data and its properties. To do this, a new query is made (Figure 10.12) that aims to show the variable that is being measured and the name of the sensor.

If you want to see what kind of short circuit occurred, you can do a simple query (Figure 10.13). You are asked to show the shorts that are biphasic to ground and their location on the map.

10.3.4.3 Geographic Representation

The query representation shown above should be in a user-friendly configuration. This involves the development of a final algorithm, which displays this data on a map. Folium is a popular Python library that allows you to create interactive maps using Leaflet.js (http://folletojs.com/). What it does, elegantly, is create Java script code that uses Brochure's wonderful interactive map library.

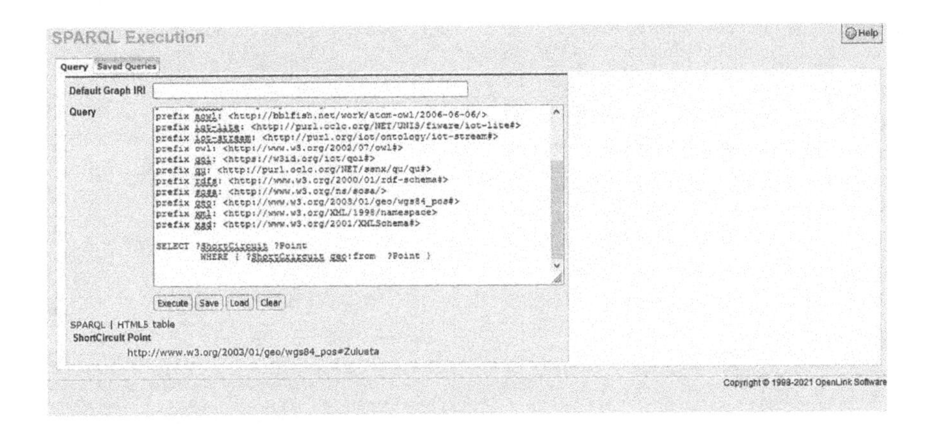

FIGURE 10.10 SPARQL query example.

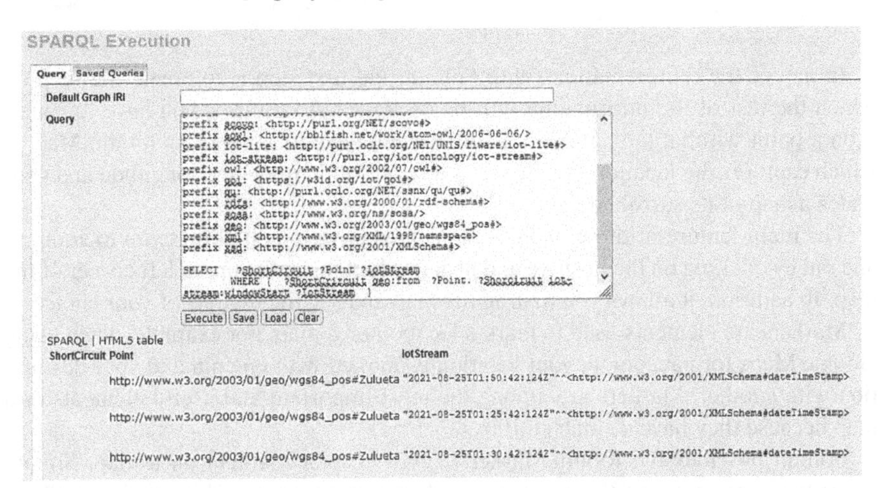

FIGURE 10.11 Consultation of faults and their geographical location.

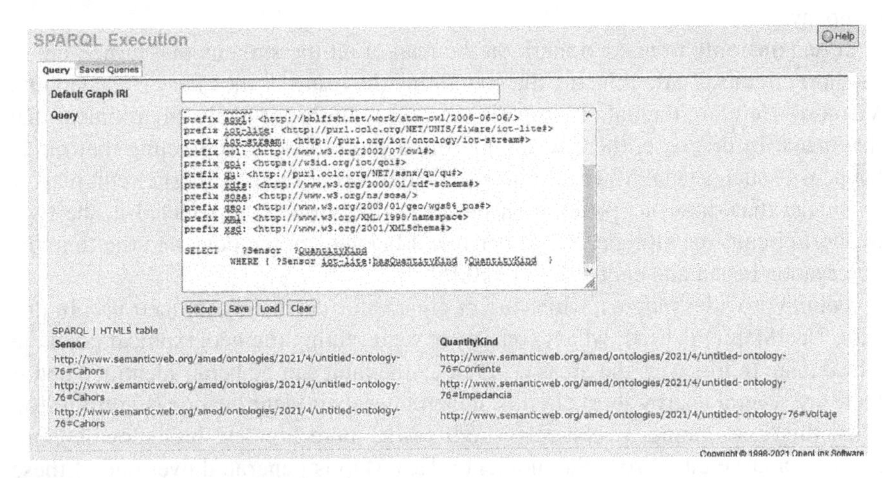

FIGURE 10.12 Consultation of sensor variables.

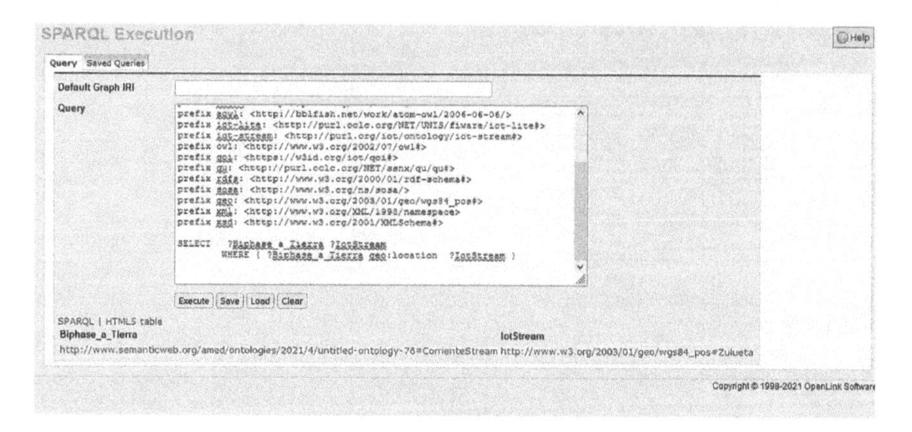

FIGURE 10.13 Search for a specific type of short circuit.

In any of the representations with Folium, the first step is to create the map on which the rest of the information will be displayed. To do this, you have to center it on a point with latitude and longitude. Folium generates the class folium.Map (), which captures the location parameter as it relates to latitude and longitude and generates a map of its surroundings.

The main feature of these maps is their interactivity, as it allows you to zoom in and out by clicking on the positive and negative buttons in the upper left corner of the map. In addition, it allows you to drag the map and see the regions of your choice.

Markers are elements used to mark a location on a map. For example, when using Google Maps for navigation, your location is marked with one pin and your destination with another. Markers are among the most important and useful elements of a map because they have so much utility.

Folium provides the folium.Marker () class to plot markers on a map. Simply insert the coordinates of this marker, which in this case is the Zulueta Power Plant. This is the reference point (Figure 10.14) of the town for the workers of the electric company.

It remains only to make a mark on the map of all the streams that turned out to be short circuits. To do this, the file containing the result of the query is loaded into Virtuoso. Here are the data to provide the user with that friendly environment that they need. In this case, there were only three short circuits, something that rarely happens in such a small town, but in larger cities it is normal. A short circuit marker is created that shows its position on the map, the sensor that detected it, the type of short circuit, the altitude, the power line it belongs to, the pole, and the time the observation began and ended (Figure 10.15).

Folium provides plugins, which are of great value depending on their use. In this case, HeatMap () is used, which consists of representing the geographical points in a heat map. In this way, the areas covered by the fault can be better identified and if there are several nearby short circuits, it facilitates their identification (Figure 10.16).

In this case, being a relatively small town, there are few breakdowns compared to a large city. From the moment a HeatMap is generated over one of these

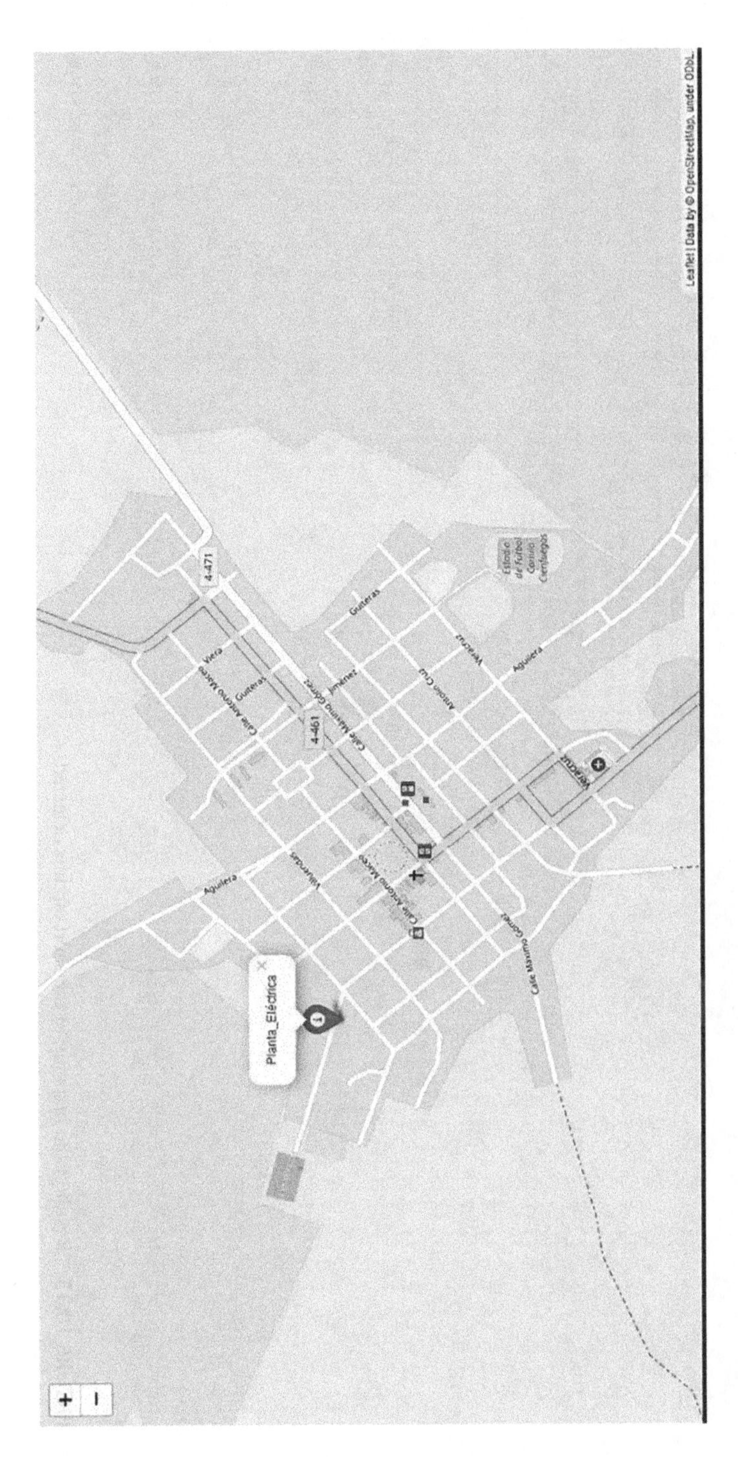

FIGURE 10.14 Map of Zulueta with a power plant marker.

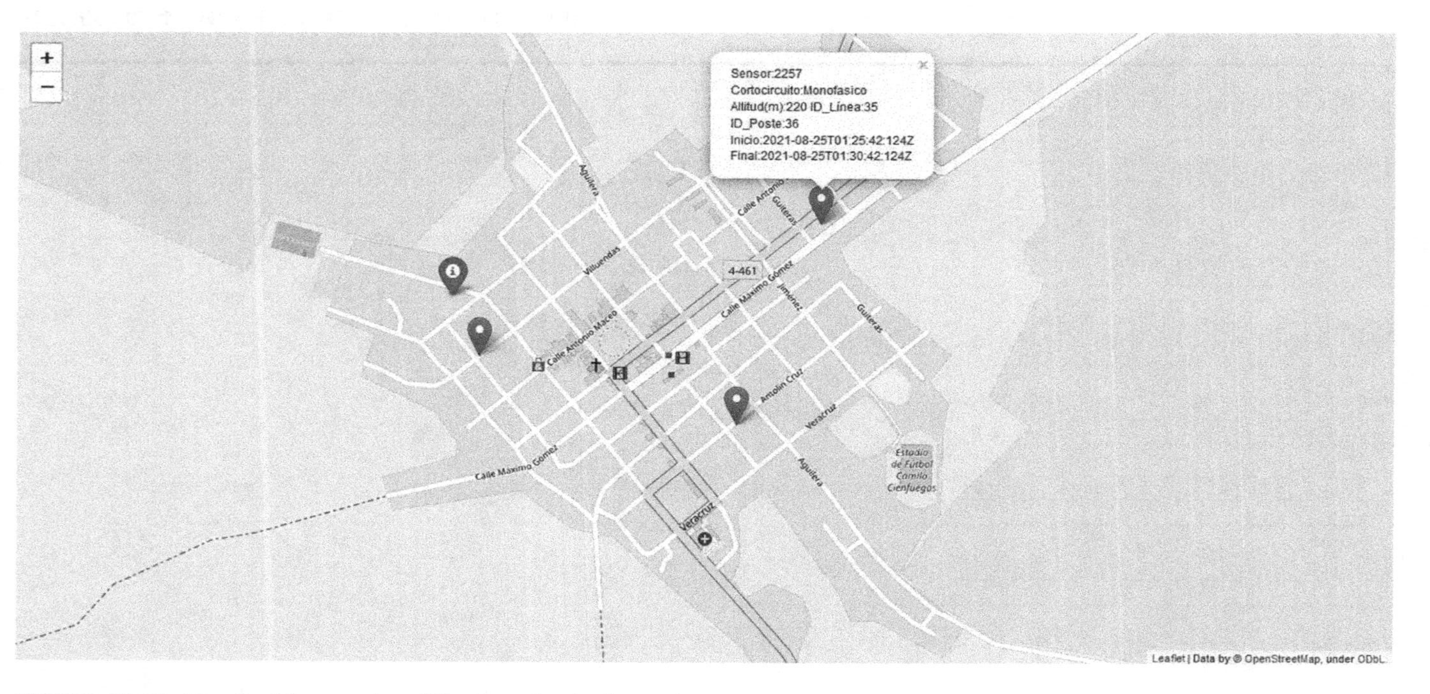

FIGURE 10.15 Result of the mapping of the short circuits detected.

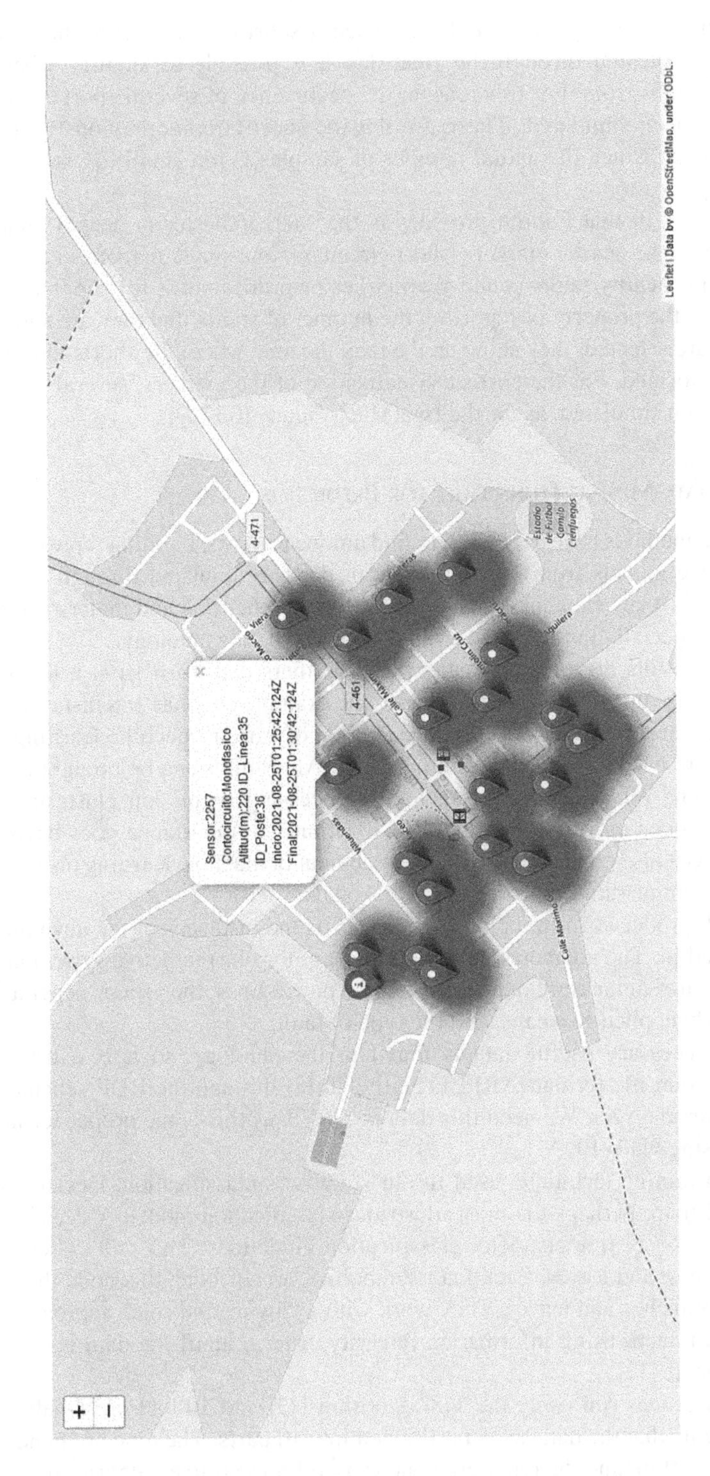

FIGURE 10.16 HeatMap.

immense cities, it can be detected that when a short circuit occurs they lead to others around it, and through the HeatMap it is possible to identify where the main fault comes from. For this reason, the occurrence of several short circuits in Zulueta has been simulated. These are not the short circuits previously inferred by the ontology, since the actual number of samples is too small to visualize the HeatMap application.

Another plugin that Folium provides is the MarkerCluster () density mapping plugin. It joins the nearby markers (short circuits) into a single marker with the aim that if several nearby faults occur, the marker map does not it fills up These new markers have the property of reporting the number of shorts that they are signaling, and if they are selected, they show on the map the area where the shorts that belong to them are located. For the correct visualization of the markers, several short circuits have been simulated, as for the HeatMap (Figure 10.17).

10.3.5 Data Mining Techniques for Prediction

Data mining methods have been used to find unique patterns, building representative, and predictive models from huge amounts of data accumulated through ontology. Data mining is a key element to achieve a better apprehension and understanding of the results, especially in the classification generated by the ontology.

The best results are obtained when the classifiers can learn from a lot of real data; in this case due to the scarcity of data, the example only has 188 instances. The selected software was WEKA,[11] is a collection of various machine learning algorithms that are used to solve data mining tasks. Algorithms can be brought directly to a dataset, and WEKA has materials for preprocessing, regression, clustering, classification, association rules, and data visualization. Its importance goes beyond its simple use, as it also contributes to the development of machine learning projects and can perform comparative research or checks on data sets.

The RDF produced from the ontology gives us the instances of the attributes we are interested in. The attributes that are important for our predictive system are the voltage and current measurements on the three power lines, the groups generated by the previously applied K-means, and the type of fault.

All the necessary useful data is found in the ontology, so it is extracted in attribute-relation file format (ARFF) (Figure 10.18) through the RDF scheme with the help of an analyzer. We need this data in ARFF, as this is the proper format for data processing via WEKA.

The data mining technique used in our study was classification. Decision trees (Figure 10.19) are perhaps the most illustrious classification model. A decision tree symbolizes a set of tree character classification guidelines. This is a collection of knots, branches, and leaves. Each knot symbolizes an attribute; this node then fragments into branches and leaves. They work with a "divide and rule" approach; each node is partitioned, using information integrity criteria, until the data is sorted to meet a halt state.[12]

Our dataset was run using the J48 algorithm [13] and 10-fold cross-validation. This represents that the data set is partitioned into 10 parts. The nine are handled to train the algorithm, and the remaining one is used for the trained algorithm.

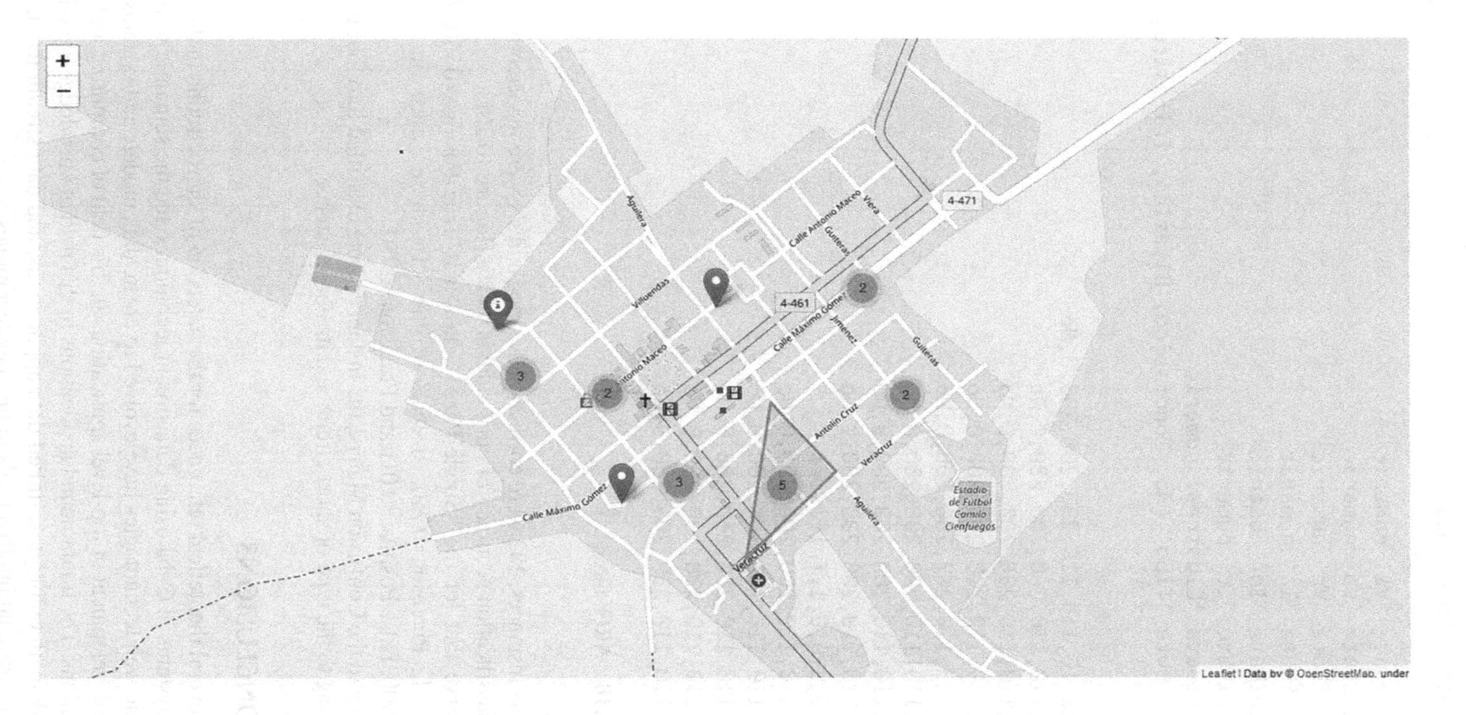

FIGURE 10.17 Various markers.

```
@relation DatosLuis

@attribute 'Va' numeric
@attribute 'Vb' numeric
@attribute 'Vc' numeric
@attribute 'Ia' numeric
@attribute 'Ib' numeric
@attribute 'Ic' numeric
@attribute 'Clusters' numeric
@attribute 'Tipo' {No, Monofasico, Bifasico, Trifasico}


@data110,110,110,200,200,200,0,No
110,110,110,200,200,200,0,No
110,110,110,200,200,200,0,No
110,110,110,200,200,200,0,No
110,110,110,200,200,200,0,No
110,110,110,200,200,200,0,No
110,110,110,200,200,200,0,No
110,110,110,200,200,200,0,No
0,0,0,0,0,0,1,Trifasico
110,110,110,200,200,200,0,No
110,110,110,200,200,200,0,No
110,110,110,200,200,200,0,No
110,110,110,200,200,200,0,No
110,110,110,200,200,200,0,No
```

FIGURE 10.18 ARFF file.

Of the 188 instances, 182 were correctly classified and six were incorrectly classified. The classification accuracy of our model is 96.8% (Figure 10.20).

We observe that for the "Single-Phase" class it is True Positive=0.750, False Positive=0.011, Precision=0.750, instead of the "Tri-Phase" class it is True Positive=1.000, False Positive=0.011 and Precision=0.750.

According to the Confusion Matrix, the instances were classified into four groups corresponding to the type of short circuit and its non-existence.

10.4 CONCLUSIONS

There are several theoretical references for the detection of short circuits in the electric power systems in Cuba using data stream techniques and the semantic web. IoT and semantic web technologies have evolved to create data models and systems that support the development of top-level applications in this field of computing and telecommunications. The fundamental applications in this field of knowledge are OBDA processes, domain ontologies, linked data applications, and data stream techniques that combine artificial intelligence and big data techniques.

```
=== Classifier model (full training set) ===

J48 pruned tree
------------------

Ia <= 0
|   Vc <= 0
|   |   Va <= 0: Trifasico (6.0)
|   |   Va > 0: Bifasico (2.0)
|   Vc > 0
|   |   Va <= 0
|   |   |   Vb <= 0: Bifasico (2.0)
|   |   |   Vb > 0: Monofasico (2.0)
|   |   Va > 0: Monofasico (6.0)
Ia > 0: No (170.0)

Number of Leaves  :     6

Size of the tree :      11


Time taken to build model: 0 seconds
```

FIGURE 10.19 Decision tree.

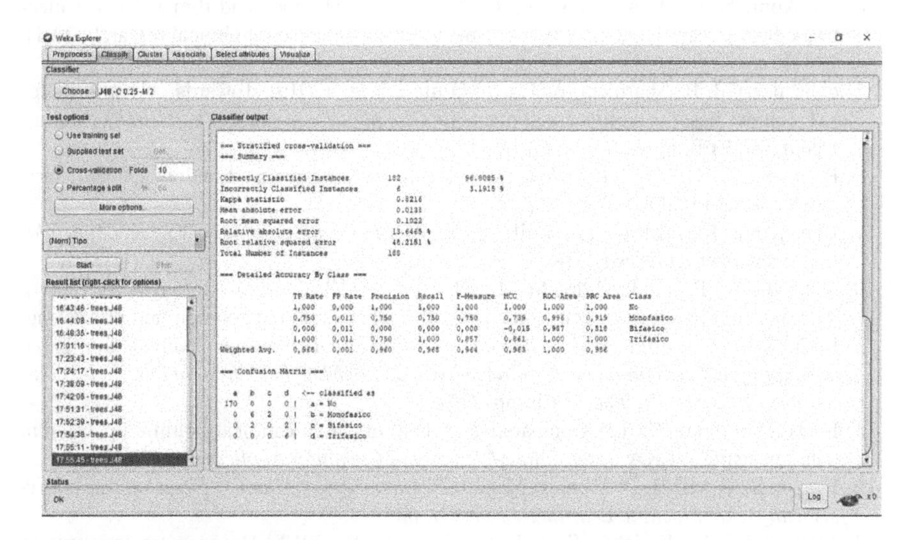

FIGURE 10.20 Algorithm J48.

The design of the short circuit detection system has all the elements used in IoT technologies for semantic web systems because it combines grouping models and incremental reasoning to facilitate decision making.

The proof of concepts demonstrates the feasibility of the design by confirming the reasoning and clustering of shorts in a given area using various visualization techniques.

NOTES

1. www.w3.org/2003/01/geo/
2. www.w3.org/TR/2017/REC-owl-time-20171019/
3. www.w3.org/2005/Incubator/ssn/ssnx/qu/qu-rec20.html

REFERENCES

1. Compton, M., et al. (2012). The W3C semantic sensor network incubator group SSN ontology. *Journal of Web Semantics*, 17, 25–32.
2. Janowicz, K., Haller, A., Cox, S. J., Le Phuoc, D., & WS Lefrançois, M. J. J. O. (2019). SOSA: A lightweight ontology for sensors, observations, samples, and actuators. *Journal of Web Semantics,* 56, 1–10.
3. Alessandro, B., Martin, B., Martin, F., Sebastian, L., & Stefan, M. (2013). *Enabling things to talk: Designing IoT solutions with the IoT architectural reference model.* Springer. https://scholar.google.com/scholar_lookup?title=Enabling%20Things%20to%20Talk%3A%20Designing%20IoT%20Solutions%20With%20the%20IoT%20Architectural%20Reference%20Model&publication_year=2013&author=B.%20Alessandro&author=B.%20Martin&author=F.%20Martin&autho
4. Battle, R., & Kolas, D. J. S. W. (2012). Enabling the geospatial semantic web with parliament and geosparql, *Semantic Web*, 3(4), 355–370.
5. Strong, D. M., Lee, Y. W., & Wang RYJC, A. (1997). Data quality in context, *Communications of the ACM*, 40(5), 103–110.
6. Weiskopf, N. G., & AMIA Weng, C. J. J. (2013). Methods and dimensions of electronic health record data quality assessment: Enabling reuse for clinical research, 20(1), 144–151.
7. Gonzalez-Gil, P., Skarmeta, A. F., & Martinez, J. A. (2019). Towards an ontology for IoT context-based security assessment. In *Global IoT summit 2019 (GIoTS).* Aarhus, Denmark: IEEE, pp. 1–6.
8. Raimond, Y. (2008). *A distributed music information system.* London: Queen Mary, University of London.
9. Belhajjame, K., et al. (2013, April 30). *Prov-o: The prov ontology: W3c recommendation. Lancaster University*, UK.
10. Kolozali, S., Bermudez-Edo, M., Puschmann, D., Ganz, F., & Barnaghi, P. (2014). A knowledge-based approach to real-time IOT data stream annotation and processing. In *2014 IEEE international conference on the internet of things (ithings), ieee green computing and communications (GreenCom), and IEEE cyber, physical and social computing (CPSCom).* Taiwan: IEEE, pp. 215–222.
11. RJIJ or. CA Arora. (2012). Comparative analysis of classification algorithms on different data sets using WEKA. *International Journal of Computer Applications*, 54(13).
12. Drazin, S., & MJML-PI Montag, M. (2012). *University of Miami, "Decision tree analysis with Weka. Machine Learning-Project II*, pp. 1–3.
13. MJIJ or. R.i. Mathuria, C. S., & Engineering, S.-W. (2013). Decision tree analysis of the j48 algorithm for data mining. *Proceedings of International Journal of Advanced Research in Computer Science and Software Engineering*, 3(6).

11 Semantic-Based Access Control for Data Resources

Ozgu Can

CONTENTS

11.1 INTRODUCTION

Access control mechanisms have an essential role to provide the protection against the increasing security and privacy vulnerabilities in today's digitalized and connected environment. Access control is a security approach to control systems' and users' interactions with resources and other systems. Thus, access control is a basic security technique to protect confidential information from unauthorized entities. As stated by Norman (2012),[1] perfect security involves perfect access control. Over the past years, access control approaches have been extensively researched and various access control solutions have been proposed to grant authorization to subjects and control access requests to resources.

The semantic web allows us to build a model of a domain and enables it to represent information more meaningfully. Thus, it provides a common and shared understanding of a domain. The semantic web technologies provide interoperability between different information systems and also improve the data quality. Besides, the semantic web empowers users to operate at abstraction levels and enables autonomous and semiautonomous agents to facilitate the collection, reasoning, and processing of data. Therefore, users conduct their processes above the technical details of format and integration.[2] In this context, an ontology, which is the core of the semantic web, has a major part in providing the interoperability. An ontology can define a semantically-rich knowledge base for the information systems and integrate information coming from different sources.

Conventional access control mechanisms only deal with the authorization decisions on the subject's access rights on a resource. Further, access to resources cannot

DOI: 10.1201/9781003310792-11

be controlled securely if the access decision does not considering the semantic interrelations between entities in the data model.[3] Besides, ignoring the semantic interrelationships among entities and making decisions based on isolated entities may result in unauthorized access and incomplete access rights for granting authorization.[4] The major advantage of the semantic aware access control model is that it considers the rich semantics of the underlying data. Also, ontologies provide the automatic establishment of security metrics by considering the explicit and reasoning information, improving the effectiveness and efficiency in security operations, and supporting analysts to obtain the related information to identify threats and security vulnerabilities.[5] Thus, semantic representation of an access control model allows the representation of an expressive and flexible access control policy, reuse and share of ontological knowledge on the access control information, and the ability to perform advanced and useful reasoning on access control policies. Thereupon, semantic web technologies are used to develop various access control mechanisms and policy languages for different domains.

The importance of access control is growing due to the move to mobile, the increasing connectedness, and the ongoing digitization process in various domains. A semantic aware access control approach enhances the expressiveness of authorization rules, considers the semantics of the data, provides a dynamic and automated access control, and improves the access control system's flexibility. This chapter aims to present an enhanced knowledge on access control concepts and contribute a more detailed understanding of access control related issues, examine the current status of semantic web-based access control solutions, and form a future direction for the research agenda of semantic aware access control mechanisms and models.

The chapter comprises the following sections: Section 11.2 clarifies the core concepts of access control and the basic access control models and mechanisms. Section 11.3 investigates the recent studies and approaches that focus on semantic web-based access control. Section 11.4 analyses the future directions of semantic aware access control. Finally, Section 11.5 presents the conclusion with a concise overview of the chapter.

11.2 ACCESS CONTROL: CONCEPTS, MODELS AND MECHANISMS

Access control is defined as "any software, hardware, procedure, or organizational administrative policy that grants or restricts access, records and monitors attempts to access, identifies users attempting to access, and determines whether access is authorized".[6] Thus, access control addresses issues more than the control action. The following subsections present the fundamental concepts, models, and mechanisms of access control in order to comprehend these issues and the basic cornerstones of access control.

11.2.1 Terminology and Basic Concepts

In information technology, access is defined as the information transfer from an *object* to a *subject*.[6] In this context, a *subject* or *principal* is an entity that seeks data or information from *objects* where data/information is stored. Eventually, subjects such as a user, program, or process are active entities, whereas objects such as a computer, data storage device, printer, file, database, program, or process are passive entities. Objects do not manipulate other objects in the system.[7] The subject's and object's roles can interchange while they interact to perform a task. For instance, when a query request is sent from a program to a database, the program is the subject that seeks information from the database. Conversely, when the database returns the query result to the program roles reverse and the database becomes the subject. [6] The key elements of a basic access control model are depicted in Figure 11.1.[8] Consequently, a source is a principal for a request, a request is an operation to be performed on an object, the guard is a reference monitor that examines each access request for the object and decides whether to grant the access request or not, and a resource is an object that is going to be accessed. An Access Control List (ACL) is an access rule that is attached to an object. The ACL defines a set of authorized principals for each operation. Then, the access request is granted by the reference monitor if its principal is trusted.[8]

Access control is a fundamental component to ensure the CIA security triad which is referred as Confidentiality, Integrity, and Availability principles. The CIA security triad model shown in Figure 11.2 forms the three-legged stool upon which the security systems are built.

In order to ensure security, these three key concepts should be addressed:

- *Confidentiality:* Confidentiality is a principle to prevent unauthorized persons from accessing the data. The goal is to prevent the disclosure of information to unauthorized access. Thus, information is only accessed by authorized entities that need to access it. The loss of confidentiality results in an unauthorized disclosure of information. It should be noted that confidentiality and privacy are two different concepts. The *privacy* term is used when data is related to personal information. Personally identifiable information (PII) is any information about the individual. PII can be used to identify the individual's name, place and date of birth, social security number, or any other information that is linkable or linked to an individual, such as financial, educational, and medical information.[9] Privacy enables individuals to control what information related to them may be gathered and stored and to whom and by whom that information may be revealed.[10]
- *Integrity:* Integrity is a principle to prohibit unauthorized modification and alteration of data. Integrity assures that data can only be modified by authorized entities. Thus, integrity prevents authorized entities from making

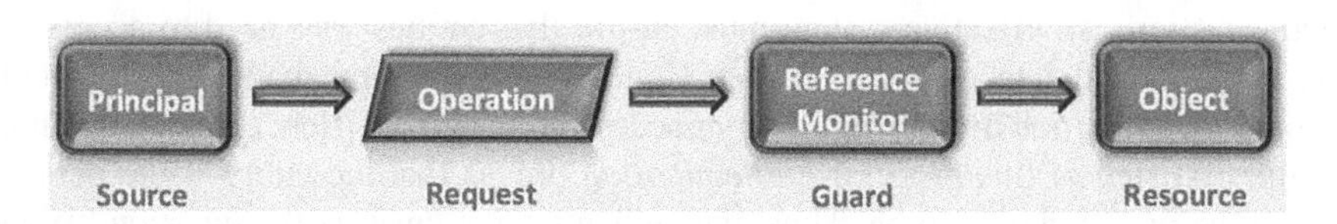

FIGURE 11.1 Access control's basic elements.[8]

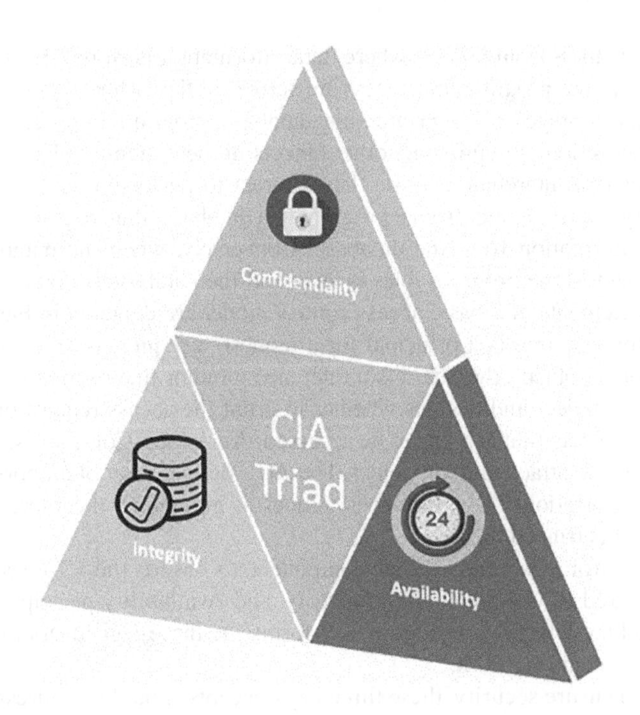

FIGURE 11.2 The CIA triad of security.

unauthorized changes to the data. Integrity covers two related concepts:[10] *data integrity* and *system integrity*. The data integrity guarantees that data can only be modified in an authorized manner. The system integrity guarantees that the system's intended function is performed in an intact manner and also ensures that no unauthorized manipulation has occurred in the system either intentionally or unintentionally. The loss of integrity results in unauthorized manipulation or unauthorized modification of information or system.

- *Availability:* Availability is a principle to ensure that information and systems are accessible to authorized entities in a reasonable amount of time upon their requests. Thus, availability assures that services are not denied to authorized users and systems work promptly.[10] The loss of availability results in interruption of access to information, a service, or a system.

Access control enforces both confidentiality and integrity. Besides these fundamental concepts of security objectives, access control has two steps: authentication and authorization. An effective system must ensure that entities who need to access to resources have the necessary privileges. Confidentiality assures that entities without proper authorization are prevented from accessing resources. Hence, authentication is the first step of this process. *Authentication* proves that the entity is the person whom he or she claims to be. Thus, the authentication answers the question "Who said this?".[8] For this purpose, authentication process binds the user ID to an entity by using passwords, smart cards, hardware/software tokens, or biometrics. When the entity is authenticated, actions that the authenticated entity can perform on a system are described by the authorization process. *Authorization* assigns privileges and grants permissions/rights to an entity that requests access to a resource and performs actions on that resource. Thus, the authorization answers the question "Who is trusted to access this?".[8] Further, the *access control mechanism* checks whether the entity has permitted to access the resource or not. The access control first checks if the entity who requests access to the resource is authenticated. Then, the access control mechanism checks the authenticated entity's privileges for the actions that the entity requests to perform. Thus, the access to information or system is controlled and limited. Also, the access request is authorized only if it does not conflict with the stated authorizations.[11] Access control systems focus on who, where, and when notions. Therefore, access is granted on a "Who, Where, and When" formula which states that each authorized user (*Who*) is valid for the related resources (*Where*) and on a certain authorized schedule (*When*).[1] Figure 11.3 shows the main components of an access control mechanism.[11] As shown in Figure 11.3, access control policies specify the high-level requirements for composing authorizations.

In brief, the process of determining if a subject is allowed to perform an action on an object is known as access control. Thus, the access control model constructs the access decision by specifying concepts and relationships between these concepts.[12] Therefore, the access control model forms the underlying model in which the security policies are defined. Thus, an access control policy is a set of rules that is built on the access control model. A policy defines concepts related with access control,

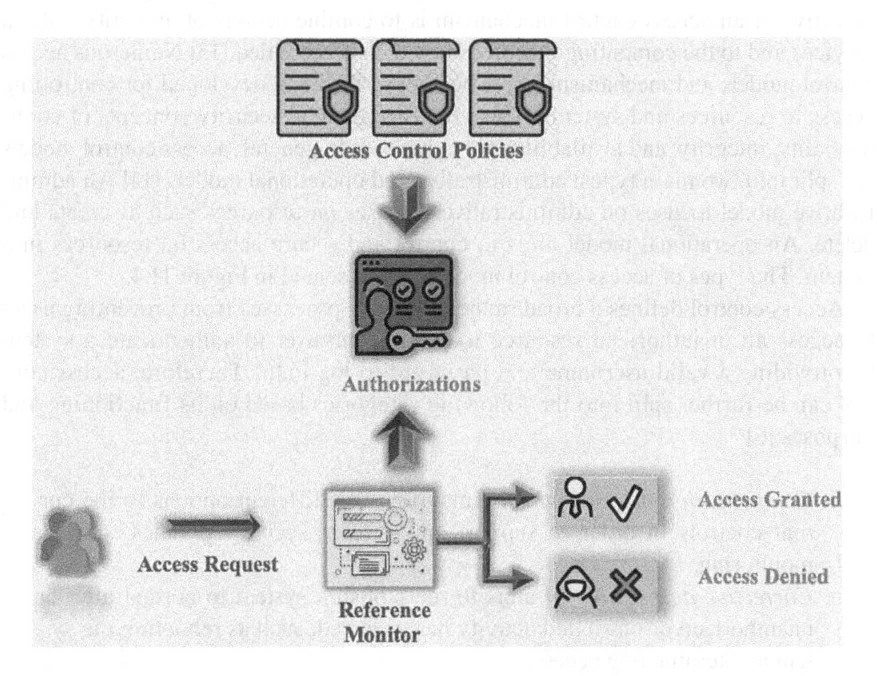

FIGURE 11.3 Main components of the access control mechanism.[11]

such as subjects, objects, permissions, and the other related concepts within a system. Finally, the access control policy determines the authorization decisions.[12]

Another important issue in an access control model is to determine how much information the subject should be permitted for access. For this purpose, Need to Know, Separation of Duties (SoD), Least Privilege, and Maximized Sharing are key principles in access control. The *Need to Know* principle states that each subject must be able to access only the information that is necessary for its purpose. The principle of *Separation of Duties* or *Segregation of Duties* assures that the critical operations are divided into tasks that are performed by more than one person. Thus, SoD aims to prevent errors and frauds in a system by restricting the assignment of responsibility to one person. The *Least Privilege* principle specifies that the subject should be authorized with the minimal/least amount of access privilege needed for the subject to perform its task. With Need to Know, the subject must need to know that specific piece of information before accessing it.[7] Moreover, the Least Privilege can restrict the Need to Know. Briefly, the Need to Know concerns specific information that the subject is working on; the Least Privilege principle restricts access rights of subjects and provides the least amount of access to the subject. The principle of *Maximized Sharing* preserves the confidentiality and provides the integrity of sensitive information by enabling the maximum sharing of information among subjects.[11]

11.2.2 Access Control Models and Mechanisms

In security, access control is a core concept and an essential principle. The basic objective of an access control mechanism is to confine actions of an entity only to services and to the computing resources to which it is entitled.[13] Numerous access control models and mechanisms have been described and developed for controlling access to resources and systems. Thus, the fundamental security concepts of confidentiality, integrity, and availability are enhanced. In general, access control models are split into two main types: administrative and operational models.[14] An administrative model focuses on administrative accesses on resources such as create and delete. An operational model aims to control and secure access on resources in a system. The types of access control models are presented in Figure 11.4.

Access control defines a broad range of control processes from preventing users to access an unauthorized resource to forcing the user to authenticate a system by providing a valid username and password to log in.[6] Therefore, access control can be further split into the following categories based on its functioning and purposes:[6]

i. *Compensation access control* aims to ensure different options to the current controls in order to support and enforce security policies, such as monitoring.
ii. *Corrective access control* aims to reestablish a system to normal after an unauthorized or unwanted activity has occurred, such as rebooting the system and terminating access.
iii. *Detective access control* aims to detect an unauthorized or unwanted activity, such as intrusion detection systems, recording, and reviewing events.

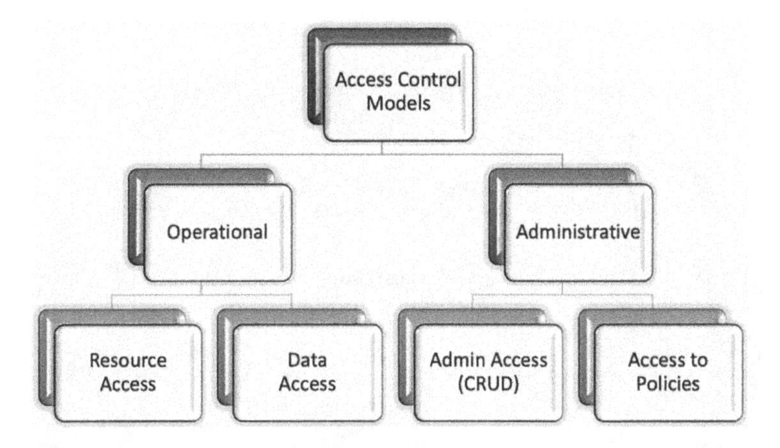

FIGURE 11.4 Types of access control models.[14]

iv. *Deterrent access control* aims to prevent the security policy violations, such as auditing and using encryption methods.
v. *Directive access control* aims to control subjects' actions to enforce compliance with security policies, such as security awareness training and posting notifications.
vi. *Preventive access control* aims to prevent unauthorized or unwanted activity from occurring, such as penetration testing and antivirus software.
vii. *Recovery access control* aims to respond to access violations of security policies by repairing the damage and restoring resources and capabilities, such as backups and fault tolerant systems.

Besides the functioning and purposes, access control is also categorized according to the type of implementation as follows:[6]

i. *Administrative access control* aims to implement the overall access control of an organization and enforce it, such as policies and procedures.
ii. *Technical/logical access control* are software or hardware models that aim to provide protection for systems and resources and also manage access to them, such as passwords, smart cards, biometrics, and firewalls.
iii. *Physical access control* aims to prevent physical access with systems, such as motion detectors, locked doors and alarms.

A layered security which is known as the defense-in-depth strategy is ensured by deploying a multiple levels/layers of access control. As shown in Figure 11.5, multiple levels/layers of defense should be overcome by attackers in order to reach the protected objects/assets.[15]

Access control solutions have been extensively researched and proposed in the literature for many years. Therefore, there are several forms of access controls, and each access control technique has its own security properties to define how subjects

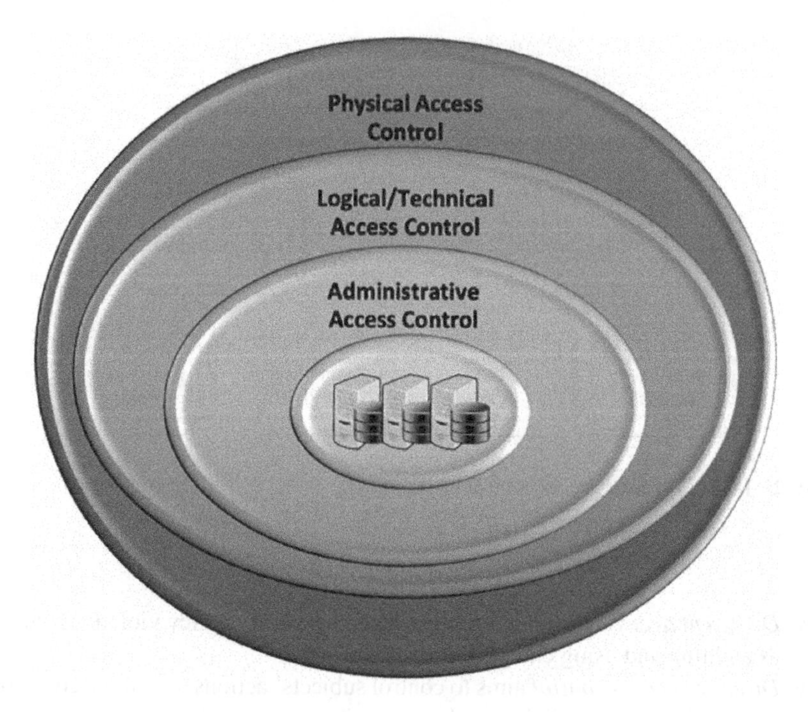

FIGURE 11.5 The layered security.[15]

access to objects and interact with these objects.[6] In this context, the main access control techniques are:

i. *Discretionary Access Control (DAC)*: The DAC permits the creator/owner of an object to control the access to that object and set the security level settings for subjects. Thus, the decision of the owner of the object indicates the access control decision. An administrative level access is not required to assign permissions.[16] DAC is often implemented using an Access Control List (ACL). ACL defines the type of access to subjects as granted or prohibited. DAC does not perform as a centrally controlled access management system.[6] For instance, if a user is the owner of a file, then the user can modify the permissions of the related file by granting or denying access to other subjects.
ii. *Mandatory Access Control (MAC)*: The MAC restricts the access to resources by using the classification labels. A subject is granted access to an object based upon the security clearance level of the subject. For this purpose, the subject's level of security clearance is used to label the subject and the object's level of classification/sensitivity is used to label the object.[6] For example, the military and government settings utilize the classifications of Top Secret, Secret, Confidential, and Unclassified. According to the MAC approach, user has no control over security settings to provide any privileges to any subject. The MAC is based on the principle of Need to Know. Administrators grant subjects to access to objects based upon their Need to Know requirements.

iii. *Role-Based Access Control (RBAC)*: In RBAC mechanism, users are assigned to the appropriate roles and permissions are associated with roles. Therefore, permissions of the security policy are given to the roles, not to the user. Thus, users are assigned to roles, permissions are assigned to roles, users can be also reassigned to new roles, and permissions can be revoked from roles. Therefore, the management of permissions is simplified.[17] The relations between core RBAC elements are shown in Figure 11.6.[12, 18] For example, roles of a hospital system can be doctor, nurse, laboratory technician, physician, patient, and administrator.
iv. *Attribute-Based Access Control (ABAC)*: The ABAC model [19] considers the environmental conditions and the assigned attributes of the subject and object to give an access decision. Thus, the ABAC model is based on attributes rather than roles to determine the access decision. Figure 11.7 shows the overview of the ABAC model.[20]
v. *Organization-Based Access Control (ORBAC)*: The ORBAC model is developed due to the problems that occurred when applying the conventional access control models such as DAC, MAC, and RBAC. ORBAC aims to overcome the limitations of the traditional access control models, handle the dynamic requirements of access control, and provide a more flexible model.[21] The organization states the organized group of subjects is the key entity in the ORBAC model.[22] Further, subject, object, and action are respectively abstracted into role, view and activity. Also, the ORBAC model allows defining policy objects as permissions, prohibitions, obligations, and recommendations. The objective of the policy objects is to control the activities performed by roles on views within an organization. There are two level in the ORBAC model: abstract (role, activity, view) and concrete (subject, action, object). Figure 11.8 shows the two levels of the ORBAC model.[21]
vi. *Purpose-Based Access Control (PBAC)*: In PBAC, the purpose of the access request is used to reach an access control decision. Therefore, a subject should specify her purpose of access together with his or her access request. A hierarchical structure based on the specialization and generalization principles is used by the PBAC model to organize purposes. Hence, the PBAC authorizes each subject for a set of purpose and grants a subject through his or her role. Finally, the authorization related with a purpose permits the role of the subject to access data with that particular purpose.[23]

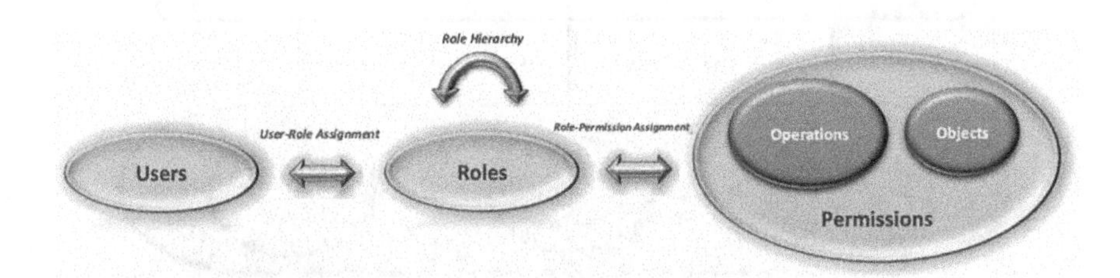

FIGURE 11.6 Relations between RBAC elements.[12, 18]

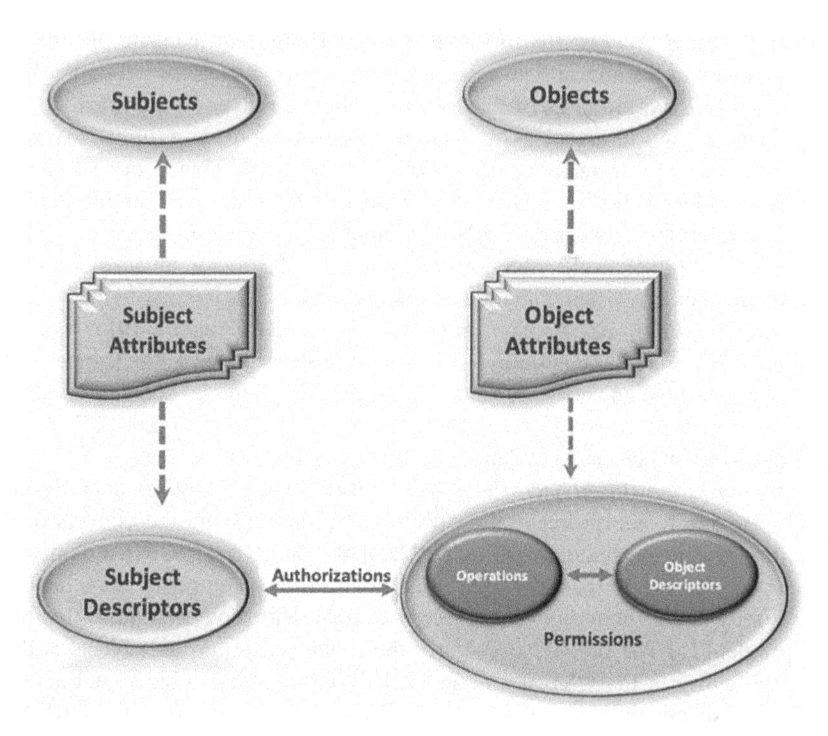

FIGURE 11.7 The overview of the ABAC model.[20]

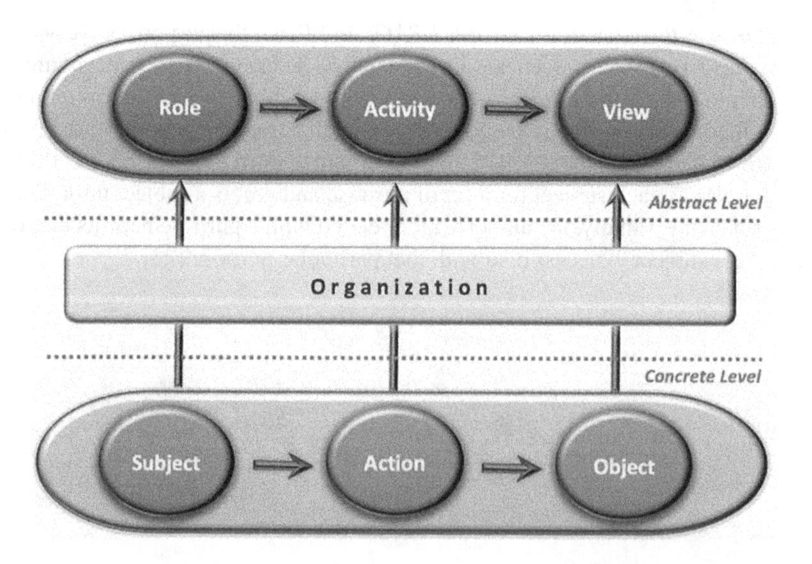

FIGURE 11.8 The ORBAC levels.[21]

vii. *Rule-Based Access Control (RuBAC)*: The RuBAC defines restrictions/rules to specify what is and is not allowed in a system. In the RuBAC model, the predefined rules are used to grant accesses in the system resource.[16] These rules define restrictions on how, when, where, and under what circumstances a subject can access a resource or a system. The model is generally combined with other access control models.

viii. *Risk-Based Access Control*: The Risk-Based Access Control is a dynamic model. A dynamic access control model relies on contextual and real-time data. For this purpose, both dynamic and real-time features and static policies are used to make access decisions.[24] Thus, the Risk-Based Access Control utilizes the changing security risk value of each access request. Hence, this risk value is used as a criterion to determine access decisions.[25] Therefore, a risk analysis on access requests is performed to make an access decision. The Risk-Adaptable Access Control (RadAC) model, which is a Risk-Based Access Control, uses a combination to authorize access privileges. The combination includes the subject's identity, the level of security risk that exists between the subject and the system being accessed, and the mission need.[26]

ix. *Time-Based Access Control (TBAC)*: The TBAC model determines the access decision based on time and date information. For this purpose, the temporal expressions (e.g. on working days between 08:00 a.m. and 17:00 p.m.) and temporal intervals (e.g. [12/2021,10/2022]) are used to specify access restrictions.[27] Therefore, the model allows to express temporal dependencies among authorizations.

In addition to these models mentioned, the literature consists of various access control models. The most commonly used models among these are Bell-LaPadula confidentiality model,[28] Biba's integrity model,[29] Chinese-Wall security policy model,[30] Clark and Wilson integrity model,[31] Temporal-RBAC,[32] and Geo-RBAC model.[33] Hence, research studies on access control models continue to develop more expressive models in line with the current technological developments and recent requirements in the field.

11.3 THE SEMANTIC WEB-BASED ACCESS CONTROL

The semantic web is an extension of the current web. Tim Berners-Lee [34] realized inadequacies of traditional web technologies in managing and sharing information. Hence, he introduced the semantic web in order to make the web more intelligent and provide a new form of web content that computers can understand. Therefore, the semantic web enables us to represent information more meaningfully, allows machine interpretable representation of concepts, and enables contents to be understandable by both machines and humans.[35, 36] From the knowledge management perspective the current web has some limitations, such as searching for information depends on keyword-based search engines, extracting information requires human effort and time, problems occur in maintaining information due to inconsistencies in outdated information and terminology, uncovering information is difficult for weakly structured and distributed documents, and viewing information is hard to realize due to the restricted access to certain information.[37]

The semantic web aims to provide advanced knowledge management systems. For this purpose, semantic web uses formal semantics to enable machines to reuse and share the information. Thus, a well-defined model of the domain information is provided, machines communicate with other machines, and the reusability is achieved. The shared machine-readable representation and the reusability is realized with ontologies. Ontology is described by Gruber (1993) as an explicit and formal specification of a shared conceptualization.[38] Thus, it provides a common and shared understanding of a domain by modeling the aspects of the domain. For this purpose, an ontology is used to describe concepts and specify interrelations between these concepts.

The semantic web is accepted as a web that is sophisticated and highly intelligent with little or no human intervention needed to accomplish tasks.[39] Besides, semantic interoperability enables different parties exchange data with each other by allowing them to access and interpret unambiguous data.[40] Also, a semantically-rich knowledge base is provided by ontologies. Therefore, the automatic establishment of security metrics is supported based on the explicit information that is represented by ontologies.[5] Furthermore, ontologies enable to describe concepts of information security domain knowledge, simplify the reasoning that is based on the interrelationships between data, support the information security risk management methodologies, and improve the effectiveness and efficiency in security operations.[41] Therefore, semantic web technologies are used to develop more effective systems to protect the security of personal data and preserve privacy. Thus, effective and efficient mechanisms are required to provide the security of semantic web-based approaches.[42] Thus, an access control mechanism indicates certain constraints that the subject must achieve before performing an action in order to ensure a secure semantic web.[43]

In semantic web-based access control, it is assured that only authorized subjects are granted to access resources in a semantic aware manner. Also, a semantic aware access control approach ensures that subjects can access only and all the information authorized to them.[44] Thus, ontologies define authorizations over concepts, and policies are used. A policy defines a set of rules for accessing a resource. Thus, policies are used to control access to a resource. Semantic web-based policy management enables it to specify rules for accessing resources and provides subjects to interpret and comply with the defined rules.[43]

The literature presents various policy languages defined for policy specification. The most well-known policy languages are eXtensible Access Control Markup Language (XACML),[45] KAoS,[46, 47] Ponder,[48] Proteus,[49] Protune,[50] Rei,[46, 51] and WSPL.[52] The traditional XACML is an XML-based markup language. Thus, XACML lack the knowledge representation that is necessary to handle computer-interpretable effects.[53] KAoS defines policy ontologies based on DAML. The KAoS Policy Ontologies (KPO) express policies of authorizations and obligations. Ponder policy language is an object-oriented language that allows to express policy management specifications for networks and distributed systems.[46] Rei is an OWL-Lite based policy specification language that adopts ontology technologies and enables developers to specify access control policies over domain-specific ontologies. There are three types of constructs in Rei: (i) meta policies, (ii) policy objects, and (iii) speech acts.[51] Rei enables users to represent and express the policy concepts of permissions, prohibitions, obligations, and dispensations.[54] Rei policy language also defines speech acts. Speech acts are primitives that enable a system to exchange obligations and rights between entities.

Ontology-based policy languages are used to express access control policies for real world implementations of semantic web services. For this purpose, ontology-based policy solutions are proposed for access control. In this context, Rein (Rei and N3) is a decentralized framework to represent policies.[55] Rein enables to reason on distributed policies in the semantic web. Rein policies are represented based on Rei policy language concepts. Rein also uses N3 rules to connect policies to each other and the web.[56] The first four layers of the semantic web stack is addressed by the Multi-Layer Semantic XACML Framework (MSACF) that is proposed in Hsu (2013).[53] As XACML lacks semantic metadata that is required to reason and inference, MSACF combines semantic web technologies with XACML and allows XACML to enable reasoning while preserving the original XACML features' usability. Another XACML based policy language is WSPL. WSPL is a policy language for web services to specify authorization, privacy, reliable messaging, application-specific service options, quality-of-protection, and quality-of-service.[52]

Proteus adopts a semantic context-aware model to specify policies. The semantic web enables the high-level description and reasoning on context and policies, whereas context-awareness enables operations on resources to be managed depending on the context visibility.[49] Therefore, Proteus combines the context-awareness and semantic web to provide the dynamic policy adaptation that depends on the context changes. Similarly, Protune (PROvisional TrUst Negotiation) is a policy language and metalanguage of the REWERSE (Reasoning on the Web with Rules and Semantics) project.[50] Protune allows to specify privacy and access control policies.

Besides ontology-based policy languages, semantic technologies-based access control frameworks are also studied in the literature. The SecurOntology approach proposed in García-Crespo et al. (2011) is a semantic-based solution to allow, reason, and validate access to resources.[57] Knowledge Access Control Policy (KACP) language model presented in Chen (2008) aims to control access in virtual enterprises. [58] Another semantic aware access control approach is integrating semantic web technologies with traditional access control models. Thus, the RBAC model is supported in OWL and relationship between the OWL and the RBAC model are studied in Finin et al. (2008).[59] Similarly, an ontology-based approach that simplifies the maintenance and specification of ABAC policies is introduced in Priebe et al. (2007).[20] Relation Based Access Control (RelBAC), proposed in Giunchiglia et al. (2009),[60] is an ontology-based approach that creates subjects and objects as lightweight ontologies and represents permissions as relations. Another ontology-based access control model introduced in He et al. (2010) combines RBAC and ABAC models to support separation of duty constraints, dynamic role assignment, and reasoning on role hierarchy.[61] Also, a Semantic Based Access Control (SBAC) model is given in Javanmardi et al. (2006).[3] SBAC authenticates users by using credentials that are offered by uses when requesting an access right.

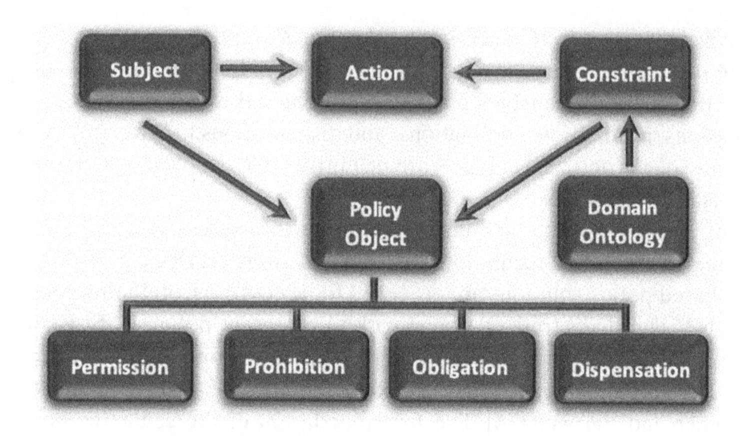

FIGURE 11.9 Policy components of OBAC.[42]

A domain-independent Ontology-Based Access Control (OBAC) that is proposed in Can et al. (2010) and Can (2009) achieves a personalized policy management by integrating domain, profile and preference information.[43, 62] The policy components of OBAC model is shown in Figure 11.9.[42] The steps in defining access control policies that are created using the related policy components are as follows:

i. Defining policy actions.
ii. Defining the constraints for each action based on the domain ontology.
iii. Specifying the type of policy object.
iv. Associating the defined policy object with the subject.
v. Storing the policy in the policy repository.

The OBAC model is based on Rei policy language. In OBAC, policy rules are formed by policy objects and four kinds of policy objects can be expressed permissions, prohibitions, obligations, and dispensations. Permission specifies actions that the subject can perform; prohibition specifies actions that the subject can't perform; obligation specifies actions that the subject should perform, and dispensation specifies actions that the subject needs no longer required to perform.

Finally, ontology-based access control is also studied for cloud computing and social networks. These researches aim to improve the performance of cloud systems and the flexibility of social network systems by addressing the privacy concerns and security issues. For this purpose, a semantic web-based approach is proposed and ontologies are used to model social network data and represent interrelations between different social network concepts.[63] Moreover, an ontology-based access control model that aims to provide protection also for the relations on a social network knowledge base is presented in Masoumzadeh and Joshi (2011).[64] In addition, ontology-based access control models to protect cloud data are proposed in Choi et al. (2014), Liu (2014), Sun et al. (2014), and Michael et al. (2019).[65–68]

Consequently, the semantic web establishes a common understanding and improves the meaning of information. Therefore, an efficient data management is provided. Hence, the literature shows that semantic web technologies are used in numerous different domains to represent much richer forms of subjects, resources, actions, and relationships among concepts. Therefore, an ontology-based approach enables to implement real world access control applications of semantic web services.

11.4 SEMANTIC AWARE ACCESS CONTROL: FUTURE DIRECTIONS

The access control mechanisms and models are needed to ensure security by protecting data and systems. The semantic aware access control mechanism ensures that only authorized subjects are granted access to resources. It should be ensured that every qualified user must be able to access each resource that he or she is authorized. Also, semantic web-based access control mechanisms describe the rich semantic interrelations in the domain model and enable the exchange of access knowledge and information between entities. Thus, a dynamic access control mechanism is provided with a semantic-aware access control model.

In the context of access control, organizations deal with challenges such as granularity, semantics awareness, consistency, efficiency, safety, and understandability.[4] Thus, the main strategic goals of semantic access control policies are stated as follows: (i) adopting a broad notion of policy, (ii) utilizing the knowledge representation and reasoning to minimize the effort required to customize general frameworks to specific application domains, (iii) integrating business rules, action specification languages, reputation models, and trust negotiation into a single policy framework, (iv) facilitating the adoption of policy languages, and (v) providing a flexible policy framework to meet the usability and efficiency requirements.[69] An ontology-based access control solution that allows the management of access authorization should feature the flexibility for security administrators to determine who can access what knowledge, how and when the use of knowledge is to be constrained, and depth and breadth of the knowledge sharing.[58] Moreover, modeling access control concepts with an ontology expands the execution of an access control mechanism and reduces the time of agreement between heterogeneous environments.[68]

Another important issue is applying access control mechanisms in compliance with regulations and standards, such as the European Union's General Data Protection Regulation (GDPR),[70] Health Insurance Portability and Accountability Act (HIPAA),[71] and California Consumer Privacy Act (CCPA). For this purpose, various consent-based access control mechanisms are developed to achieve security and preserve privacy. Consent-based access control systems grant access requests based on the obtained consents from subjects. Semantic aware consent-based access control models that are proposed in Can (2013), Olca and Can (2016), Olca and Can (2018) integrates access control and personalization techniques.[72–74]

Provenance-based access control is also a prominent issue in data security and privacy. Provenance data determines the origin of the data. For this purpose, operations performed on data are traced and recorded.[35, 36] A fine-grained, comprehensive, and flexible provenance-based access control mechanisms that consider the security and privacy requirements are needed.[75] For this purpose, ontology-based provenance approach is proposed in Can and Yilmazer (2020), Can and Yilmazer (2020), Can and Yilmazer (2014), and Yilmazer and Can (2021) to protect data from unwanted access and detect access violations.[35, 36, 76, 77]

Further, a generalized conceptual access control model is also an essential requirement for a common access control mechanism. A generalized ontology-based access control approach enables to model the entities involved in access control system. In addition, delegation of access control rights and resolving policy conflicts that are generated by inappropriate policy specifications are important issues in the field. Also, more fine-grained access control mechanisms are necessary to represent complex relations.[64]

Finally, access control policies must handle access decisions under a broad range of performance requirements and risk levels.[69] The semantic web describes the evolution from a document-based web towards a new paradigm.[57] This paradigm represents information more meaningfully and facilitates knowledge sharing and reuse. Thus, semantic technologies-based access control solutions provide the flexibility to effectively address the security issues and privacy concerns. Further, the semantic web-based solution approaches maintain better understanding and control on the access control policies.

11.5 CONCLUSION

The semantic web provides an efficient data management by representing the comprehensive meaning of information. For this purpose, formal semantics is used. Therefore, semantic web enables machines to communicate with other machines. An ontology which is the core of the semantic web allows to specify the conceptualization of the domain information. Also, interrelations between the domain data can be determined by ontologies. Therefore, an ontology can represent the formal specification of a shared conceptualization for an authorization policy. Thus, adding semantics to access control models and developing semantic aware security models are prolific research lines.[57] For this purpose, an access control approach should consider the semantics of the data. Hence, new access control models and mechanisms are required to enable users to define rights by specifying the semantic relationships between various objects.[4]

Access control is a significant and a challenging problem to ensure confidentiality and integrity. Semantic web technologies are used to develop various policy languages and access control models for different domains. Besides, the semantic web enables the expression of meaningful access control rules and addresses the changing dynamics of subjects, objects, and access grants.[60] Therefore, controlling access to resources can be effectively achieved and subjects can be authorized through an ontology-based approach. Thereby, semantic web technologies enable the implementation of more secure and privacy-aware access control systems. Moreover, the semantic web enables to create flexible and expansible access control models and architectures. Consequently, semantic web-based solutions are still open for future research in implementing access control and authorization models. The recent developments in today's digitalized and connected environment present new challenges for access control studies.

REFERENCES

1. Norman, T. (2012). *Electronic access control*. Watham, MA: Butterworth-Heinemann.
2. Compton, M., Barnaghi, P., Bermudez, García-Castro, R., Corcho, O., Cox, S., Graybeal, J., Hauswirth, M., Henson, C., Herzog, A., Huang, V., Janowiczk, K., Kelsey, W. D., Le Phuoc, D., Lefort, L., Leggieri, M., Neuhaus, H., Nikolov, A., . . .Taylor, K. (2012). The SSN ontology of the W3C semantic sensor network incubator group. *Web Semantics: Science, Services and Agents on the World Wide Web*, 17, 25–32.
3. Javanmardi, S., Amini, M., & Jalili, R. (2006). An access control model for protecting semantic web resources. In *Proceedings of the ISWC'06 workshop on semantic web policy (SWPW'06), CEUR workshop proceedings*. vol. 207, Athens, Georgia: CEUR-WS.
4. Ryutov, T., Kichkaylo, T., & Neches, R. (2009). Access control policies for semantic networks. In *IEEE international symposium on policies for distributed systems and networks*. Piscataway, NJ: IEEE, pp. 150–157.
5. Mozzaquatro, B. A., Agostinho, C., Goncalves, D., Martins, J., & Jardim-Goncalves, R. (2018). An ontology-based cybersecurity framework for the internet of things. *Sensors*, 18(9), 3053.
6. Stewart, J. M., Tittel, E., & Chapple, M. (2011). *CISSP, certified information systems security professional study guide* (Fifth Edition). Indianapolis, IN: Wiley Publishing.
7. Conrad, E., Misenar, S., & Feldman, J. (2017). *Eleventh hour CISSP* (Third Edition). Cambridge, MA: Syngress.
8. Lampson, B., Martin, A., Burrows, M., & Wobber, E. (1992). Authentication in distributed systems: Theory and practice. *ACM Transactions on Computer Systems*, 10(4), 265–310.
9. McCallister, E., Grance, T., & Scarfone, K. (2010). *Guide to protecting the confidentiality of personally identifiable information (PII)*. Gaithersburg, MD: NIST Special Publication, pp. 800–122.
10. Stallings, W., & Brown, L. (2018). *Computer security: Principle and practice* (Fourth Edition). New York, NY: Pearson.
11. Ferrari, E. (2010). *Access control in data management systems*. Williston, VT: Morgan & Claypool Publishers.
12. Thion, R. (2008). *Access control models. Cyber warfare and cyber terrorism*. Hershey, PA: IGI Global, Chapter 37, pp. 318–326.
13. Benantar, M. (2006). *Access control systems—security, identity management and trust models*. New York: Springer.
14. Gupta, M., Bhatt, S., Alshehri, A. H., & Sandhu, R. (2022). *Access control models and architectures for IoT and cyber physical systems*. Switzerland: Springer.
15. Chapple, M., Stewart, J. M., & Gibson, D. (2018). *(ISC)² CISS certified information systems security professional official study guide* (Eighth Edition). Indianapolis: John Wiley & Sons.
16. Meyers, M., & Jernigan, S. (2018). *Mike Meyers' CompTIA security+ certification guide* (Second Edition). New York, NY: McGraw-Hill Education.
17. Sandhu, R. S., Coyne, E. J., Hal, L. F., & Youman, C. E. (1996). Role-based access control models. *IEEE Computer*, 29(2), 38–47.
18. Ferraiolo, D. F., Kuhn, D. R., & Chandramouli, R. (2007). *Role-based access control* (Second Edition). Norwood, MA: Artech House Publishers.
19. Hu, V. C., Ferraiolo, D., Kuhn, R., Schnitzer, A., Sandlin, K., Miller, R., & Scarfone, K. (2014). *Guide to attribute based access control (ABAC) definition and considerations* (draft). Gaithersburg, MD: NIST Special Publication, 800, no. 162 (2013): 1–54.
20. Priebe, T., Dobmeier, W., Schläger, C., & Kamprath, N. (2007). Supporting attribute-based access control in authorization and authentication infrastructures with ontologies. *Journal of Software (JSW)*, 2(1), 27–38.

21. Cuppens, F., & Miège, A. (2003). Modelling contexts in the Or-BAC model. In *19th annual computer security applications conference*. Piscataway, NJ: IEEE, pp. 416–425.
22. Kalam, A. A. E., El Baida, R., Balbiani, P., Benferhat, S., Cuppens, F., Deswarte, Y., Miege, A., Saurel, C., & Trouessin, G. (2003). Organization based access control. In *Proceedings of POLICY 2003: IEEE 4th international workshop on policies for distributed systems and networks*. Piscataway, NJ: IEEE, pp. 120–131.
23. Byun, J.-W., Bertino, E., & Li, N. (2005). Purpose based access control of complex data for privacy protection. In *Proceedings of the 10th ACM Symposium on Access Control Models and Technologies (SACMAT '05)*, New York, NY: ACM, pp. 102–110.
24. Atlam, H. F., Azad, M. A., Alassafi, M. O., Alshdadi, A. A., & Alenezi, A. (2020). Risk-based access control model: A systematic literature review. *Future Internet*, 12(6), 103.
25. Dos Santos, D. R., Westphall, C. M., & Westphall, C. B. (2014). A dynamic risk-based access control architecture for cloud computing. In *2014 IEEE network operations and management symposium (NOMS)*. Piscataway, NJ: IEEE, pp. 1–9.
26. Singhal, A., Winograd, T., & Scarfone, K. (2007). *Guide to secure web services*. Gaithersburg, MD: NIST Special Publication, pp. 800–895.
27. Samarati, P., & di Vimercati, S. C. (2001). Access control: Policies, models, and mechanisms. In *Foundations of security analysis and design (FOSAD 2000)*, vol. 2171. Berlin, Heidelberg: Springer, pp. 137–196.
28. Bell, D. E., & LaPadula, L. J. (1973). *Secure computer systems: Mathematical foundations*. Springfield, VA: The Mitre Corporation.
29. Biba, K. J. (1977). *Integrity considerations for secure computer systems*. Bedford, MA: The Mitre Corporation.
30. Brewer, D. F. C., & Nash, M. J. (1989). The Chinese wall security policy. In *Proceedings of the 1989 IEEE symposium on security and privacy*. Piscataway, NJ: IEEE, pp. 206–214.
31. Clark, D. D., & Wilson, D. R. (1987). A comparison of commercial and military computer security policies. In *1987 IEEE symposium on security and privacy*. Piscataway, NJ: IEEE, pp. 184–184.
32. Bertino, E., Bonatti, B. A., & Ferrari, E. (2001). TRBAC: A temporal role-based access control model. *ACM Transactions on Information and System Security*, 4(3), 191–233.
33. Damiani, M. L., Bertino, E., Catania, B., & Perlasca, P. (2007). GEO-RBAC: A spatially aware RBAC. *ACM Transactions on Information and System Security*, 10(1), 1, Article 2, 42 pages.
34. Berners-Lee, T., Hendler, J, & Lassila, O. (2001). The semantic web. *Scientific American Magazine*, pp. 29–37.
35. Can, O., & Yilmazer, D. (2020). A novel approach to provenance management for privacy preservation. *Journal of Information Science*, 46(2), 147–160.
36. Can, O., & Yilmazer, D. (2020). Improving privacy in health care with an ontology-based provenance management system. *Expert Systems*, 37(1), e12427.
37. Antoniou, G., & van Harmelen, F. (2004). *A semantic web primer*. Cambridge, MA: The MIT Press.
38. Gruber, T. R. (1993). A translation approach to portable ontologies. *Knowledge Acquisition*, 5(2), 199–220.
39. Thuraisingham, B. (2008). *Building trustworthy semantic webs*. Northwestern: Auerbach Publications, Taylor & Francis Group.
40. Rhayem, A., Mhiri, M. B. A., & Gargouri, F. (2020). Semantic web technologies for the internet of things: Systematic literature review. *Internet of Things*, 11, 100206.
41. Mavroeidis, V., & Bromander, S. (2017, September 11–13). Cyber threat intelligence model: An evaluation of taxonomies, sharing standards, and ontologies within cyber threat intelligence. In *Proceedings of the 2017 european intelligence and security informatics conference (EISIC)*. Piscataway, NJ: IEEE, pp. 91–98.

42. Can, O., & Unalir, M. O. (2010). Ontology based access control. *Pamukkale University Journal of Engineering Sciences*, 16(2), 197–206.
43. Can, O., Bursa, O., & Unalir, M. O. (2010). Personalizable ontology based access control. *Gazi University Journal of Science*, 23(4), 465–474.
44. Qin, L., & Atluri, V. (2003). Concept-level access control for the semantic web. In *ACM workshop on XML security*. New York, NY: ACM, pp. 94–103.
45. OASIS. (2013). *eXtensible access control markup language (XACML) Version 3.0*. OASIS Standard. https://docs.oasis-open.org/xacml/3.0/xacml-3.0-core-spec-os-en.html
46. Tonti, G., Bradshaw, J. M., Jeffers, R., Monranari, R., Suri, N., & Uszok, A. (2003). Semantic web languages for policy representation and reasoning: A comparison of KaoS, Rei, and Ponder. In *The 2nd international semantic web conference (ISWC 2003)*. Berlin, Heidelberg: Springer, pp. 419–437.
47. Uszok, A., Bradshaw, J. M., & Jeffers, R. (2004). KAoS: A policy and domain services framework for grid computing and semantic web services. In *The second international conference on trust management (iTrust 2004)*, vol. 2995. LNCS, pp. 16–26.
48. Damianou, N., Dulay, N., Lupu, E., & Sloman, M. (2001). The Ponder policy specification language. In *Proceedings of workshop on policies for distributed systems and networks (POLICY 2001)*, vol. 1995. Springer-Verlag, LNCS, pp. 18–38.
49. Toninelli, A., Montanari, R., Kagal, L., & Lassila, O. (2007, June 13–15). Proteus: A semantic context-aware adaptive policy model. In *Proceedings of the eighth IEEE international workshop on policies for distributed systems and networks (POLICY '07)*. Piscataway, NJ: IEEE, pp. 129–140.
50. Bonatti, P. A., & Olmedilla, D. (2005). *Policy language specification*. Technical report, Working Group I2, REWERSE project, EU FP6 Network of Excellence (NoE).
51. Kagal, L., Finin, T., & Joshi, A. (2003). A policy language for a pervasive computing environment. In *Proceedings of the 4th IEEE international workshop on policies for distributed systems and networks (POLICY'03)*. Piscataway, NJ: IEEE, pp. 63–74.
52. Anderson, A. H. (2004). An introduction to the web services policy language (WSPL). In *Proceedings of the fifth IEEE international workshop on policies for distributed systems and networks (POLICY 2004)*. Piscataway, NJ: IEEE, pp. 189–192.
53. Hsu, I. C. (2013). Extensible access control markup language integrated with semantic web technologies. *Information Sciences*, 238, 33–51.
54. Kagal, L. (2002). *Rei: A policy language for the me-centric project*. Palo Alto, CA: HP Laboratories.
55. Kagal, K., Berners-Lee, T., Connolly, D., & Weitzner, D. J. (2006). Using semantic web technologies for policy management on the web. In *Proceedings of the 21st national conference on artificial intelligence (AAAI 2006)*. vol. 2. Palo Alto, CA: American Association for Artificial Intelligence Press, pp. 1337–1344.
56. Rein. The Rein policy framework for the semantic web. https://groups.csail.mit.edu/dig/2005/05/rein
57. García-Crespo, A., Gómez-Berbís, J. M., Colomo-Palacios, R., & Alor-Hernández, G. (2011). SecurOntology: A semantic web access control framework. *Computer Standards & Interfaces*, 33, 42–49.
58. Chen, T.-Y. (2008). Knowledge sharing in virtual enterprises via an ontology-based access control approach. *Computers in Industry*, 59, 502–519.
59. Finin, T., Joshi, A., Kagal, L., Niu, J., Sandhu, R., Winsborough, W. H., & Thuraisingham, B. (2008). ROWLBAC—representing role based access control in OWL. In *Proceedings of the 13th ACM symposium on access control models and technologies (SACMAT '08)*. New York, NY: ACM, pp. 73–82.
60. Giunchiglia, F., Zhang, R., & Crispo, B. (2009). Ontology driven community access control. In *Proceedings of the ESWC2009 workshop on trust and privacy on the social and semantic web (SPOT2009)*. Heraklion, Greece, CEUR Workshop Proceedings, vol. 447.

61. He, Z., Huang, K., Wu, L., Li, H., & Lai, H. (2010). Using semantic web techniques to implement access control for web service. In *International conference on information computing and applications (ICICA 2010)*. vol. 105. Berlin, Heidelberg: Springer, pp. 258–266.
62. Can, O. (2009). Personalizable ontology based access control for semantic web and policy management. Doctoral dissertation, Ege University. Ege University, Graduate School of Natural and Applied Science. https://tez.yok.gov.tr/UlusalTezMerkezi/TezGoster?key=NtBAevXNhYaNqJFoAcdBdqU7ocOMnntKldgg3Rf9qzLJALpQtWmI1RARYBr-whQG
63. Carminati, B., Ferrari, E., Heatherly, R., Kantarcioglu, M., & Thuraisingham, B. (2011). Semantic web-based social network access control. *Computers & Security*, 30, 108–115.
64. Masoumzadeh, A., & Joshi, J. (2011). Ontology-based access control for social network systems. *International Journal of Information Privacy, Security and Integrity*, 1(1), 59–78.
65. Choi, C., Choi, J., & Kim, P. (2014). Ontology-based access control model for security policy reasoning in cloud computing. *The Journal of Supercomputing*, 67, 711–722.
66. Liu, C. L. (2014). Cloud service access control system based on ontologies. *Advances in Engineering Software*, 69, 26–36.
67. Sun, H., Zhang, X., & Gu, C. (2014). Role-based access control using ontology in cloud storage. *International Journal of Grid and Distribution Computing*, 7(3), 1–12.
68. Michael, A., Kothandaraman, R., & Kaliyan, K. (2019). Providing ontology-based access control for cloud data by exploiting subsumption property among domains of access control. *International Journal of Intelligent Engineering and Systems*, 12(3), 280–291.
69. Bonatti, P. A., Duma, C., Fuchs, N., Nejdl, W., Olmedilla, D., Peer, J., & Shahmehri, N. (2006). Semantic web policies—a discussion of requirements and research issues. In *The semantic web: Research and applications (ESWC 2006)*, vol. 4011. Lecture Notes in Computer Science, pp. 712–724.
70. GDPR. (2016). General data protection regulation. Regulation (EU) 2016/679 of The European Parliament and of The Council of 27 April 2016.
71. HIPAA. Health Information Privacy. (2022). www.hhs.gov/hipaa/index.html
72. Can, O. (2013). A semantic model for personal consent management. In *The 7th metadata and semantics research (MTSR 2013)*. vol. 390, Berlin, Heidelberg: Springer, pp. 146–151.
73. Olca, E., & Can, O. (2016). A meta-consent model for personalized data privacy. In *The 10th international conference on metadata and semantics research (MTSR 2016)*. Berlin, Heidelberg: Springer.
74. Olca, E., & Can, O. (2018). Extending FOAF and relationship ontologies with consent ontology. In *The 3rd international conference on computer science and engineering (UBMK2018)*. Piscataway, NJ: IEEE, pp. 542–546.
75. Ni, Q., Xu, S., Bertino, E., Sandhu, R., & Han, W. (2009). An access control language for a general provenance model. In *Workshop on secure data management (SDM 2009)*. vol. 5776. Berlin, Heidelberg: Springer, pp. 68–88.
76. Can, O., & Yilmazer, D. (2014). A privacy-aware semantic model for provenance management. In *The 8th metadata and semantics research conference (MTSR 2014), CCIS 478*. Berlin, Heidelberg: Springer, pp. 162–169.
77. Yilmazer Demirel, D., & Can, O. (2021). Enriching the open provenance model for a privacy-aware provenance management. *European Journal of Science and Technology*, 29, 144–149.

12 Ontological Engineering
Research Directions and Real-Life Applications

Archana Patel and Narayan C. Debnath

CONTENTS

12.1 INTRODUCTION

A machine will behave intelligently if the underlying representation scheme exhibits knowledge that can be achieved by representing semantics. Knowledge representation is the heart of artificial intelligence (AI)[1] that aims to make a machine intelligent. Good knowledge representation schemes or formalisms encode knowledge, beliefs, actions, feelings, goals, desires, preferences, and all other mental states in the machine. Various knowledge representation formalisms have been spawned; each formalism has its own set of features and tradeoffs. These formalisms differ from each other in the way that knowledge is compiled, the extent of the descriptions they offer, and the type of inference power that they sanction. Some of the better-known formalisms for knowledge representation are semantic networks, frame systems, rules, and logic.[2] Ontology is a world-famous and widely used logic-based knowledge representation formalism. Ontology is a means of representing semantic knowledge and includes at least classes (C), properties (P), relations (R), instances (I), and axioms (A).[3] The key requirement of ontology is the development of suitable languages for the representation and extraction of information. Varieties of ontology languages have been developed, and the most operable and standard language is Web

DOI: 10.1201/9781003310792-12

Ontology Language (OWL).[4–7] Ontology query language plays a very important role in extracting and processing information. SPARQL is one of the most widely used ontology query languages.[8] By using these semantic technologies, users and systems can interact and share information with each other in an intelligent manner.

Ontology engineering is a sub-field of knowledge engineering that provides a set of methods and methodologies for the development of ontologies.[9] The aim of the ontology engineering field is to provide a direction for solving the problem of interoperability via ontology. With the help of ontology, we can offer the personalized response of the user queries and mitigate the risk of the stock market. In the real world, the domain knowledge, as well as the contextual information, are somewhat imperfect, meaning that the imprecision and uncertainty are not represented yet are demanding for a personalized response. The major problem is how to provide contextualized and personalized behavior in the real world that is intrinsically uncertain. This problem becomes a challenge when the user has several different constraints on the available resources. Certainty and specificity are just like two faces of a coin; certainty is inversely proportional to specificity. Computers are bound by the logic of 0 and 1; they cannot derive new knowledge based on stored facts. Say it is asked, "Where is Michael?" The answer may be

- Michael is in the city Mumbai; if asked to be very quick.
- Michael is working outdoors in city Mumbai; if asked to be specific but may be less certain.
- Michael is planting crops in city Mumbai; if asked to be highly specific.

Humans decide upon a course of action based on past experience, gathered knowledge, and current scenario. In contrast, machines try every possible combination but are not able to figure out a perfect answer in context to the problem at hand. It befits us, in designing automated systems, to represent an inference with such imperfect knowledge.[10] The main problem is to provide reasoning under resource constraints with imprecise and uncertain information. Variable Precision Logic (VPL) [11] provides an approach to vary the precision of conclusions in order to produce the most accurate answer possible within constraints. VPL systems may be based on variable certainty or variable specificity or both of them. In variable certainty systems, the more time is given, the more certain are the answers. In variable specificity systems, the more time is given, the more specific are the answers. Both the approaches should be combined to reflect the fact that specificity and certainty are the two opposite sides of a coin. One of them has to be sacrificed to achieve the other.

Precision refers to the closeness of two or more results to each other when the results are obtained under the same conditions (user). If you are asked to identify an animal, for example, a tiger, and your answer

- is tiger each time, then the answer is 100% precise and 100% accurate
- is cat each time, then the answer is 100% precise but less accurate because though both animals are the same species, both are not the same
- is rat each time, then the answer is, again, 100% precise but not accurate
- varies each time, then the answer is less precise, and accuracy is based on whether the answer is correct each time.

Ontology is becoming the communication means for formalization of knowledge and efficient and effective exchange of information. It defines the structure of knowledge and provides semantic access to the stored data. In this chapter, we provide ten research directions of the ontology and its two applications. The first application uses the idea of knowledge unit and diagnoses the daily life queries and to show the variation in precision based on the user context (available resources, certainty, and specificity) by applying a semi-automatic approach. The confidence of the result varies based on the precision requirements of the user for the same query posed but in a different context. The second applications utilize an ontology to mitigate the risk of the stock market.

The rest of the chapter is organized as follows: Section 12.2 explains the future possible research directions of ontology. Section 12.3 shows two applications of ontology. Section 12.4 concludes this chapter.

12.2 ONTOLOGICAL ENGINEERING: RESEARCH DIRECTIONS

Ontologies envisage enormous benefits in the information technology world, and it is widely used in different fields.[12] A number of researches are going on in different aspects of ontology. The most promising research directions for the ontology are:

- Ontology representation as knowledge unit: Ontology is a semantic data model that stores real-world entities in the form of classes, properties (data and object properties), and instances. However, a standard unit is required to quantitatively measure the entities, for example, a liter is a standard unit that is used to measure liquid. Similarly, a unit is also required to encode the knowledge. The existing semantic data models do not provide a unit to encode the knowledge.
- Ontology provenance: The provenance of a resource is a record that defines the process and entities involved in producing and delivering or influencing that resource. The credibility criteria that mean 'to find the origin of the data' lead to the provenance of data. Ontologies are used to track the provenance of data as they represent data in the graph format (subject-object-predicate) and connect all data. SPARQL query language is used to extract the information from the ontologies. Data provenance is the task of the top layer of the semantic web architecture and has a lot of scope for research work.
- Ontology development from unstructured data: A lot of data is available on the web in different formats like audio, video, word files, excel sheets, PDF files, etc. The problem is how to convert this large unstructured data into an ontology. The literature shows that a schema-level ontology has been developed from unstructured data; however, there is a lot of scope for research in the context of determining ontological properties (data and object) from unstructured data.
- Ontology reasoner: Reasoner extracts the information from the ontology, and it also checks the ontology's consistency. Different profiles of OWL are available that vary from each other based on expressiveness, soundness, completeness, reasoning method, incremental classification, etc. A critical estimation and evaluation of requirements is needed before selecting a

reasoner for a real-life application. There is a need to develop an improved optimized reasoner that works with large complex ontologies (A-Box and T-Box) in a reasonable time.

- Ontology mapping: Mapping determines the relationship between the concepts of the two ontologies. The relationship can be a subset, superset, equivalence, or disjoint. The query results are extracted from multiple ontologies based on the mapping results. How to determine the correct mapping between the concepts of two ontologies within a reasonable time is a challenge. Nowadays, multilingual ontology mapping is a favorite topic for many researchers. OAEI provides a global platform for this task.
- Ontology merging: Merging is required in many places, like when one company wants to acquire another, they need to merge the knowledge bases. The base of merging is the mapping because ontologies or knowledge bases are merged based on the mapping results. Despite having many algorithms of matching, it is still a challenge to get a coherent ontology after merging base ontologies. Multilingual ontology merging is a favorite topic for many researchers, and OAEI provides a global platform for this task.
- Ontology partitioning: Ontology partitioning is required when we deal with large-scale ontologies. It is a crucial phase of ontology mapping/matching. Only a few matching systems are available that match the concepts of the ontology after partitioning. The major problem of ontology partitioning is to provide algorithms that make coherent partitions of the ontology and preserve all relationships in their original form.
- Ontology evaluation: Evaluation of the ontology checks the quality of the ontology. Various criteria have been proposed to determine the quality of the ontology. However, there is no concrete method/framework/tool available

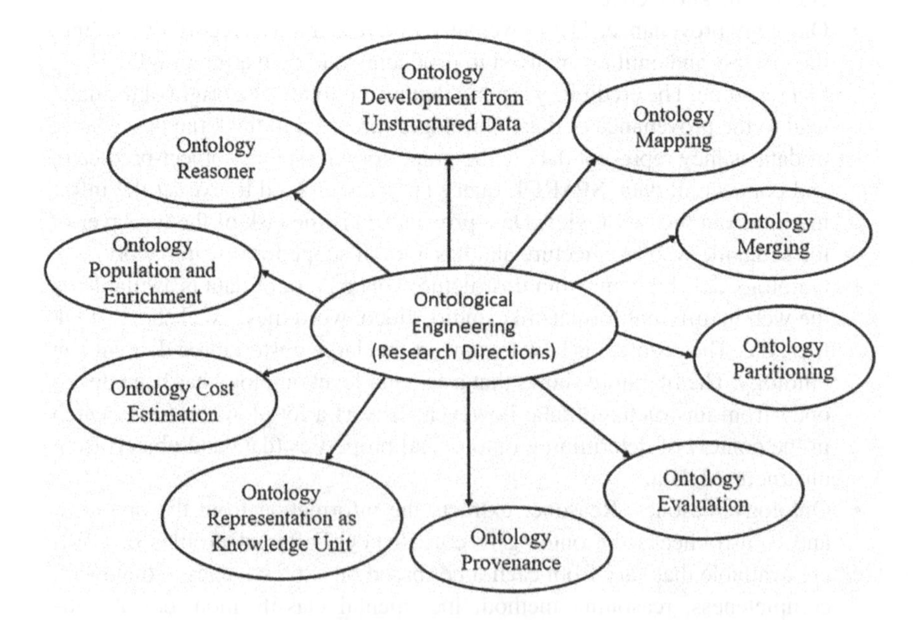

FIGURE 12.1 Research directions: Ontological engineering.

that offers a suitable ontology as per the needs of the user. There is a need to develop an effective framework to accommodate all aspects of ontology evaluation.

- Ontology cost estimation: Calculating the cost of ontology development is a very interesting topic among researchers. Only a few models are available in the literature for the same task. Recently, fuzzy methods are utilized for cost estimation. However, the sophisticated model is still missing that can be utilized to estimate the cost of the ontology.
- Ontology population and enrichment: Ontology population means to assign or add new instances of classes to the ontology, whereas ontology enrichment extends the available model of ontology by adding classes and semantic relationships. Both tasks are very challenging because they need models/algorithms that match (syntactically and structurally) the entities correctly.

12.3 APPLICATIONS OF ONTOLOGY

This section explains two applications of the ontology. One application is based on the concept of a knowledge unit to find out the precise answer for daily life human queries in an interactive way in accordance with the imposed constraints and user choice of certainty and specificity. Another application shows the ontological framework for risk mitigation in the stock market.

12.3.1 Personalized Behaviour

We have developed an ontology called normal routine (NR) ontology based on the knowledge unit as advocated by Patel et al. (2018) and then run on a semi-automatic approach called Diagnostic Belief Algorithm (DBA) proposed by authors Jain and Patel (2019) to show the precise answer of the imposed query according to the choice of the certainty and specificity.[13-14]

12.3.1.1 Ontological Concept as Unit of Knowledge (UoK)

When users develop ontologies, they do not concentrate on representing every entity as a whole. Most of the time, some properties that are important to match the concept are last in the queue, hence misleading the results. This section explains the representation of knowledge as a unit. An entity/concept must be represented as a unit of knowledge (UoK) with all the important dimensions specified. Jain and Patel (2019) have defined a knowledge unit that can be utilized to describe any entity/concept in its entirety.[14]

$$\mathrm{D}[\mathrm{TE}, \mathrm{AE}, \mathrm{VE}, \mathrm{PE}](\omega) = \langle DP(\gamma), \mathrm{CP}, \mathrm{C}(\delta), \mathrm{G}, \mathrm{S}, \mathrm{I} \rangle$$

A detailed view of this tuple is presented in Figure 12.2.

The entity/decision/concept/event D is considered to be true (with confidence value ω) if

- The m distinctive features DP relegated with the If_{def} part of the rule are all true (Anding). The other set of properties CP relegated with the If_{charac} part

D [TE, AE, VE, PE] ω		*{Decision/Concept/Event}*
If_{def}	DP$\{DP_1{:}\omega_1; DP_2{:}\omega_2; \ldots DP_m{:}\omega_m\}\ \gamma$	*{Distinctive Features (ANDing)}*
If_{charac}	CP $\{CP_1{:}P_1C_1; CP_2{:}P_2C_3;\ldots,CP_p{:}P_pC_1\}$	*{Cancellable Features}*
Gen	G	*{General concept}*
Spec	S $\{S_1,S_2, \ldots, S_k\}$	*{Specific concepts (XORing)}*
Unless	C $\{C_1{:}\delta_1; C_2{:}\delta_2; \ldots, C_n{:}\delta_n, UNK\},\delta$	*{Exceptions to the rule (ORing)}*
Instances	I [Temporal Details, Spatial Details]	*{Instances}*

FIGURE 12.2 Concept in ontology.

of the rule enumerates the set of p cancellable features associated with the concept D. The CPs are cancellable meaning that an instance of a concept may override its value or not even possess this property.

- It belongs to the general concept G mentioned in the Gen operator.
- None of the exceptions C mentioned with the Unless operator are true. C_s are checked if time permits based on the control protocol as mentioned in Jain and Patel (2019).[14]

The set of distinctive features of a concept should be complete and mutually exclusive with the set of distinctive features of all its siblings. P_1C_1 is the default value of property CP_1 for the concept D chosen from the constraints list of property CP_1 $\{P_1C_1, P_1C_2 \ldots\}$. S operator refers to the specificity part of the rule. It enlists the set of k concepts specific to concept D in the taxonomy of ontology. For a detailed discussion of all the operators and parameters that make up the knowledge unit, the reader is motivated to read the text by Jain and Patel (2019).[14]

During matching, only distinctive features are required to match because these properties uniquely define the concepts. All the distinctive features are stored by using annotation properties. Both classical matching and realistic matching (using distinctive features) may produce the same relationship. But the internal processing for matching entities is different. In the case of classical ontology matching, every aspect of the source entity is required to be matched with every aspect of the target entity.[15] If some distinguishing property is missed during the matching (because maybe it is last in the list and time does not permit), then classical matching generates wrong results. In realistic ontology matching, only distinctive features that define/distinguish a concept from the other are subject to be matched. The cancellable features are not required to match because they do not uniquely define the concept. This type of representation reduces the existing challenges of knowledge representation and leads us to forms of representation that we believe are more natural and comprehensible than other forms.

Figure 12.3 shows a way to encode knowledge unit in the ontology. We have created an ontology that contains 10 classes and 27 properties, out of which 12 are distinctive features. Figure 12.3(a) and Figure 12.3(b) show all the class and their values of distinctive features. The distinctive features can be an object property or data type property. Annotation property has been used for the storage of distinctive features. We have created hasDefiningProperties by using annotation. All the distinctive

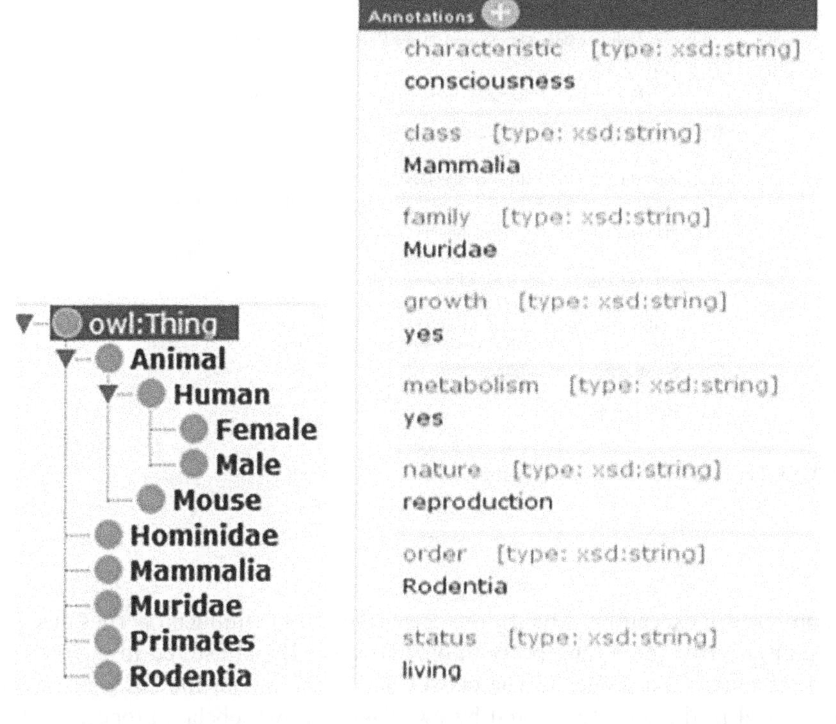

FIGURE 12.3 (a) Ontology with concept, (b) storage of distinctive features of mouse class.

features are subproperty of hasDefiningProperties. If the value of defining property is class, then select a string from the data type. If the value of defining property is the data value, then according to its value, select data type like if the value is 2 then data type should be int. Figure 12.3(b) shows the storage of value of distinctive features for mouse. The characteristics, growth, metabolism, nature, and status are the inherited distinctive features from the Animal class. The value of distinctive features, namely mammalia, muridae, and rodentia, are classes. Cancellable properties of the class have been stored using the object properties [16] and data properties.[17] A cancellable object property is defined as an instance of the built-in OWL class owl:ObjectProperty, whereas cancellable datatype property is defined as an instance of the built-in OWL class owl:DatatypeProperty. Both owl:ObjectProperty and owl:-DatatypeProperty are subclasses of the RDF class rdf:Property.

12.3.1.2 Development of Normal Routine (NR) Ontology

The NR ontology is developed by using the Protégé tool and encoded into RDF/XML format. The hierarchical structure of NR ontology is shown in Figure 12.4 where every concept is represented as a whole. The value of operator γ defines the 0-degree of strength, the value of unless operator δ defines 1-degree of strength in

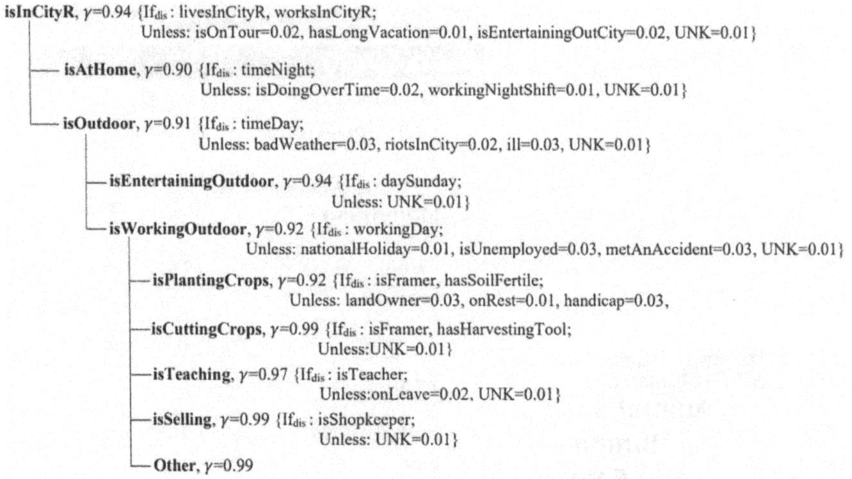

isInCityR, γ=0.94 {If_{dis} : livesInCityR, worksInCityR;
Unless: isOnTour=0.02, hasLongVacation=0.01, isEntertainingOutCity=0.02, UNK=0.01}

— **isAtHome**, γ=0.90 {If_{dis} : timeNight;
Unless: isDoingOverTime=0.02, workingNightShift=0.01, UNK=0.01}

— **isOutdoor**, γ=0.91 {If_{dis} : timeDay;
Unless: badWeather=0.03, riotsInCity=0.02, ill=0.03, UNK=0.01}

— **isEntertainingOutdoor**, γ=0.94 {If_{dis} : daySunday;
Unless: UNK=0.01}

— **isWorkingOutdoor**, γ=0.92 {If_{dis} : workingDay;
Unless: nationalHoliday=0.01, isUnemployed=0.03, metAnAccident=0.03, UNK=0.01}

— **isPlantingCrops**, γ=0.92 {If_{dis} : isFramer, hasSoilFertile;
Unless: landOwner=0.03, onRest=0.01, handicap=0.03,

— **isCuttingCrops**, γ=0.99 {If_{dis} : isFramer, hasHarvestingTool;
Unless:UNK=0.01}

— **isTeaching**, γ=0.97 {If_{dis} : isTeacher;
Unless:onLeave=0.02, UNK=0.01}

— **isSelling**, γ=0.99 {If_{dis} : isShopkeeper;
Unless: UNK=0.01}

— **Other**, γ=0.99

FIGURE 12.4 Normal routine ontology.

the presence of exceptions. The If_{dis} represents the set of defining properties that need to be always true for a concept. Both operators and If_{dis} are stored in the ontology by using annotation property. The classes and subclasses of the NR ontology are highlighted in the bold and stored by owl:class and owl:subclass properties of the ontology respectively. The UNK mentioned in the list of exceptions (with unless operator) for every concept caters to incomplete information about the concept in the real world.

12.3.1.2.1 A Semi-Automatic Approach: DBA

We have run a semi-automatic approach called the Diagnostic Belief Algorithm (DBA) over NR ontology with five type of different users, namely very low priority (VLP), low priority (LP), medium priority (MP), high priority (HP), and very high priority (VHP) user. The value of parameters threshold (m), resources (e), certainty vs. specificity (k) is assigned according to the authors Bharadwaj and Jain (1992) and Patel and Jain (2021).[18, 19] To calculate the total number of levels (N) that needs to traversed out of the total nodes (N_m) of the tree and the number of exceptions (X) that should be checked out of total present exceptions (X_m) at that node; we have utilized formulas given by the authors Patel et al. (2018).[13]

$$N = ceiling\left(N_m * \left((1.0 - m) \uparrow \left(\frac{1.0}{e * k}\right)\right)\right) \quad (12.1)$$

$$X = Round\left(X_m * \left(1.0 - (1.0 - m) * \left(\left(\frac{\mathrm{CL}+1}{\mathrm{N}}\right) \uparrow \left(\frac{\mathrm{e}}{\mathrm{k}}\right)\right)\right)\right), 0 \le \mathrm{CL} < \mathrm{N} \quad (12.2)$$

The final decision (ω_D) called 2-degree of strength is calculated from the following formula

$$\omega_D(i) = \min(\omega_D(i-1), \omega_{P1}, \omega_{P2}, \ldots \omega_{Pm}) \times \delta_i \quad (12.3)$$

Where, $\delta = \gamma + Summation of all \delta's$

If any exception is found to be true at any point during the reasoning process, that node returns false. The procedure of DBA is described below:

1. Calculate N as the level to traverse down the hierarchy of ontology based upon the user's current context by using formula 1.
2. Insert the root node into the queue.
3. Loop until a good node is found or end of queue is reached.
   a. Fetch the next node from the queue.
   b. If the values of all If_{dis} of the node are greater than a threshold for distinctive features
      Then go to step 5
      Else loop
4. End of queue is reached, so terminate by printing the contents of the stack.
5. The good node of this level is found, so empty the queue.
6. Calculate X of the good node using formula 2.
7. Calculate ω of good node using formula 3.
8. Put this good node along with its ω value into the stack.
9. Put all the children of this current good node into the queue.
10. Go to step 3.

Consider the user query, "Where is Michael?" According to the type of the user, answer will change as depicted in Table 12.1.

12.3.1.3 Findings and Discussion

There are specifically two findings that are described below.

12.3.1.3.1 Specificity (N) vs. Certainty (C)

The different aspects of precision are the certainty of belief, the specificity of conclusion, the available resources, and the allowed threshold. Certainty and specificity are inversely proportional to each other. When a user demands a highly specific answer then the certainty of that answer automatically reduces. For example: if the query is "Where is Robert?", then the more certain and less specific answer of this query is "Robert is in city Mumbai" but less certain and highly specific answer is "Robert planting crops in city Mumbai". Table 12.2 shows the variation in number of levels explored (N) and 2-level strength of implication ω for different combination of m, e, and k. Three major results that are inferred are displayed in Table 12.2 and are listed as follows:

- When the certainty increases, then the specificity decreases automatically; $\omega \propto \frac{1}{N}$

TABLE 12.1
Results for Different Context of User

Users	Different Values of m, e, k, N, X	Final Output
VLP	m=0.3, e=0.125, k=0.125, N=1, X1=1	Michael is in city Mumbai with confidence 0.8544
	m=0.3, e=0.125, k=1, N=1, X1=1(T)	Michael is not in city Mumbai with confidence 1
	m=0.3, e=0.125, k=4, N=2, X1=2, X2=1	Michael is in city Mumbai with confidence 0.803136
LP	m=0.4, e=0.6, k=0.125, N=1, X1=1	Michael is in city Mumbai with confidence 0.8544
	m=0.4, e=0.6, k=1, N=2, X1=2, X2=1(T)	Michael is in city Mumbai with confidence 0.8633
	m=0.4, e=0.6, k=4, N=4, X1=2, X2=1, X3=1, X4=1	Michael is planting crops, working outdoor in city Mumbai with confidence 0.716962017
MP	m=0.5, e=1, k=0.125, N=1, X1=2	Michael is in city Mumbai with confidence 0.8633
	m=0.5, e=1, k=1, N=2, X1=2, X2=2	Michael is outdoor in city Mumbai with confidence 0.828768
	m=0.5, e=1, k=4, N=4, X1=2, X2=2, X3=2, X4=2(F, T)	Michael is outdoor in city Mumbai with confidence 0.828768
HP	m=0.6, e=2, k=0.125, N=1, X=2	Michael is in city Mumbai with confidence 0.8544
	m=0.6, e=2, k=1, N=3, X1=3, X2=2, X3=2	Michael working outdoor in city Mumbai with confidence 0.81202176
	m=0.6, e=2, k=4, N=4, X1=2, X2=2, X3=2, X4=2 (T)	Michael working outdoor in city Mumbai with confidence 0.79561728
VHP	m=0.6, e=2, k=0.125, N=1, X=2	Michael is in city Mumbai with confidence 0.8544
	m=0.6, e=2, k=1, N=3, X1=3, X2=2, X3=2	Michael working outdoor in city Mumbai with confidence 0.81202176
	m=0.6, e=2, k=4, N=4, X1=2, X2=2, X3=2, X4=2 (T)	Michael working outdoor in city Mumbai with confidence 0.79561728

TABLE 12.2
Relationship between Certainty and Specificity for Different Values of <m, e, k>

S. No.	<m, e, k>	ω	N
1	<0.3, 0.5, 4>	0.709570656	4
2	<0.35, 0.8, 4>	0.716962017	4
3	<0.4, 1.2, 4>	0.732216528	4
4	<0.5, 0.8, 4>	0.7637925888	4
5	<0.6, 3, 4>	0.7795408896	4
6	<0.7, 2, 2>	0.81202176	3
7	<0.8, 6, 0.5>	0.83739744	3
8	<0.8, 8, 0.25>	0.845856	2
9	<0.45, 0.5, 0.25>	0.8544	1
10	<0.8, 3, 0.25>	0.8633	1

- Irrespective of m and e; $k \propto \frac{1}{\omega}$
- Irrespective of m and e; $k \propto N$

12.3.1.3.2 Variation in Precision

The maximum, minimum, and deviation of precision obtained by all five types of users are listed in Table 12.3. The deviation of precision is calculated according to Equation 4. Table 12.3 describes two major findings related to m.

$$\text{Deviation in precision} = \text{max precision—min precision} \quad (12.4)$$

TABLE 12.3
Obtained Precision

Priority of User	M	Maximum	Minimum	Deviation
VLP	0.3	0.8544	0.709570656	0.144829344
LP	0.4	0.8544	0.709570656	0.144829344
MP	0.5	0.8633	0.7637925888	0.0995074112
HP	0.6	0.8633	0.7795408896	0.0837591104
VHP	0.8	0.8633	0.8290234656	0.0342765344

TABLE 12.4
Variation in Precision for m = 0.75

e ↓/k →	k = 0.25	k = 0.5	k = 1	k = 2	k = 4
6	0.845856	0.83739744	0.8290234656	0.8290234656	0.8290234656
7	0.845856	0.83739744	0.8290234656	0.8290234656	0.8290234656
8	0.845856	0.83739744	0.8290234656	0.8290234656	0.8290234656

If m increases, then deviation in precision decreases monotonically. So, low priority users achieve high variation in precision whereas high priority users achieve less variation in precision. The high value of m generates the same precision for different value of e and k. For example, m = 0.75 generates same precision for different value of e (6,7,8) and k (0.25, 0.5, 1, 2, 4) as depicted in Table 12.4.

If m is high, then accuracy will be high. The low priority users achieve maximum 85.44% accuracy and minimum 70.95% accuracy, whereas high priority users achieve maximum 86.33% accuracy and minimum 82.90% accuracy. Accuracy will never be 1 because we deal with incomplete and uncertain information.

The uncertainty in information varies according to available resources and type of choice for certainty against specificity. Figure 12.5 depicts variation in precision for all five types of users.

12.3.2 Ontological Framework for Risk Mitigation

To mitigate the risk in stock market, various indicators have been developed by many analysts and researchers. However, no indicator alone can provide the full assurance of protecting the capital. Therefore, it is required either to develop a new indicator or propose a framework that consists of a set of those indicators that can mitigate

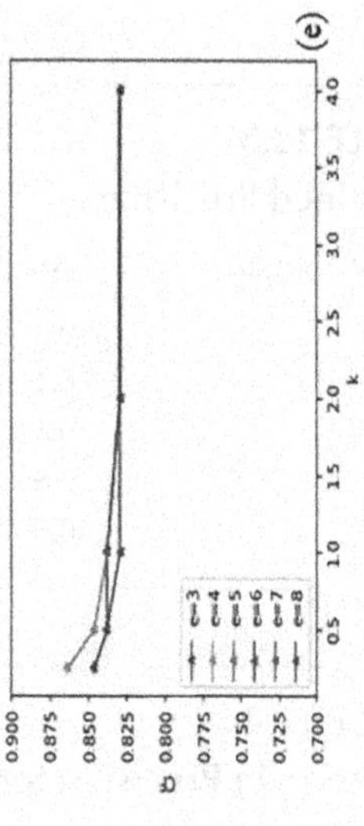

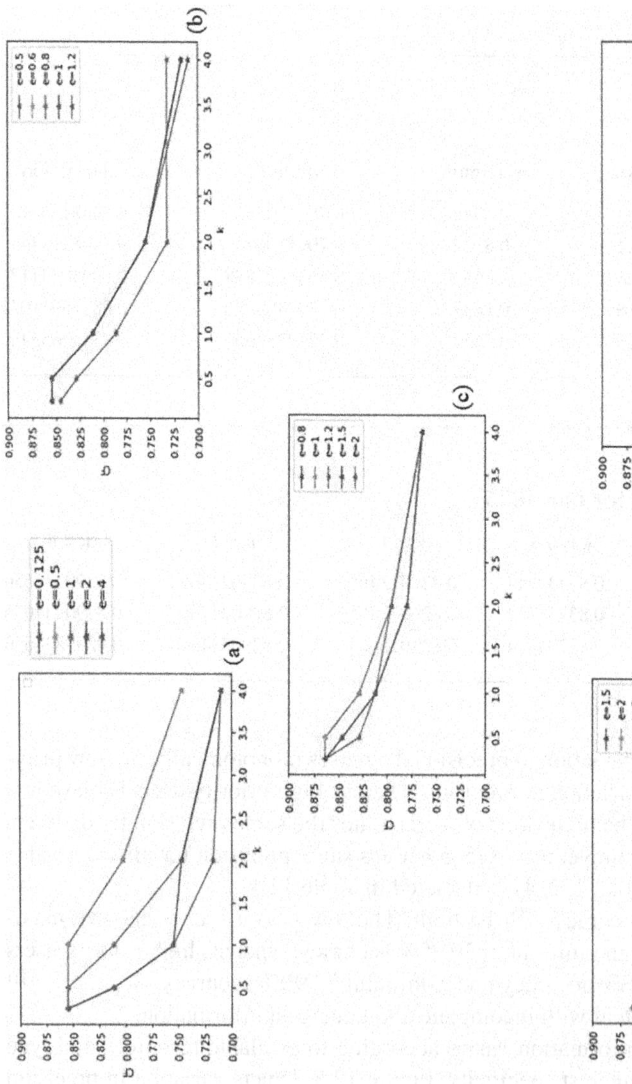

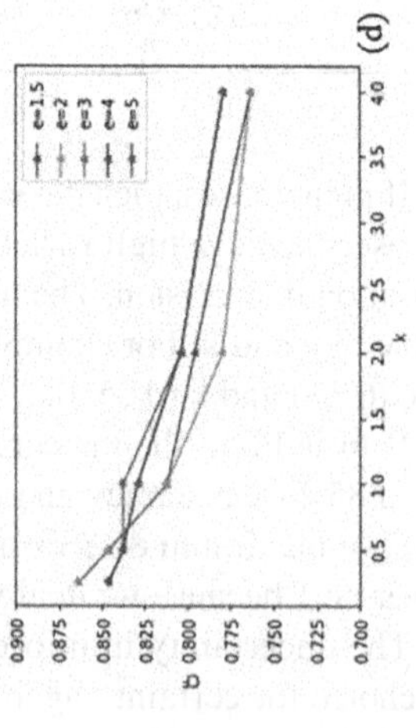

FIGURE 12.5 Variation in precision (a) VLP, (b) LP, (c) MP, (d) HP, (e) VHP.

the investors risk as compared to any of the available indicators. Here, we propose an ontology-based framework to mitigate the risk and generate greater returns. Figure 12.6 shows the working of the proposed framework which start from the technical analysis of any stock or index. This framework is an attempt to directly corelate the technical analysis with the help of chart pattern, volume, and price action with the help of some useful indicators. Thus, this framework is able to indicate the buying or selling opportunity more accurately as compared to any existing single indicator.

All the stocks of the companies are stored in the stock exchange database (we have taken BSE and NSE). The first task is to find out the appropriate stocks from the database based on the technical analysis and delivery-based volume of the stock. This technical analysis includes only stocks having two parameters namely

1. Delivery-based volume greater than 50% for the last five consecutive days.
2. Chart of stock should follow any particular pattern stored in the data base of ontological risk mitigation indicator (ORMI).

The ORMI database contains various patterns like parallel channel pattern, ascending triangle pattern, cup and holder pattern, head and shoulder pattern, double top

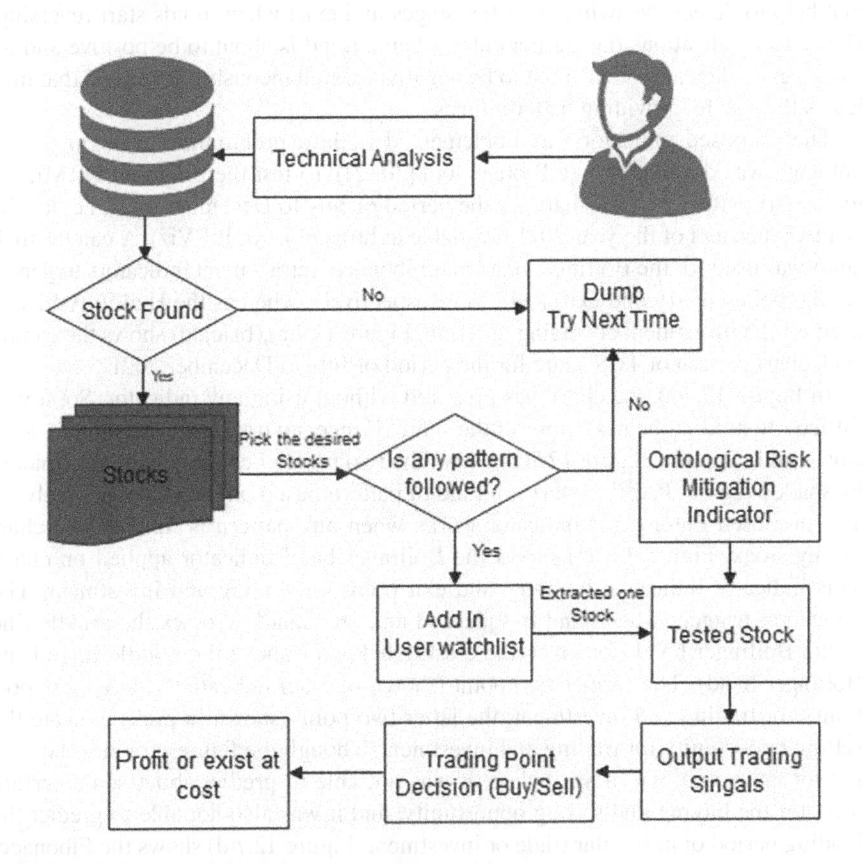

FIGURE 12.6 Ontological risk mitigation framework.

and double bottom pattern, etc. If user did not find desired stocks, dump the process and try next time. However, if a user has any stock as per above mentioned criteria, then add in the watch list. Now, pick one tested stock from the watchlist and run ORMI on the tested stock. The ORMI provides output trading signals from which the user finds trading point decision (buying and selling). Now, the user either can get profit or exist at cost.

The process takes the features of Bollinger bands, Fibonacci retracement, and Heikin-Ashi and stores all the features in the ontology for efficient search and query as the stock market is a place where numerous data is generated every second. This data is highly volatile and user needs access to it in a microsecond. This data is not only used in real time but also historically. This is the place where ontology can resolve the problems of storage and access of this highly volatile and large data on a real basis by interconnecting all the world markets. Ontology stores the extracted features of Bollinger bands, Fibonacci retracement, and Heikin-Ashi in the form of Classes (C), Properties (P), and Instances (I). The ORMI algorithm starts to work when any of the pattern (that stored in database) is followed by the chart. The proposed algorithm work better because it considers essential aspects of three indicators, namely Bollinger bands, Fibonacci retracement, and Heikin-Ashi altogether, that help to detect the swing at earlier stages and exist when trends start reversing. Hence two indications (i.e. earlier entry when a trend is about to be positive and an earlier exit when a trend is about to be negative) simultaneously trigger and that mitigates the risk by providing better returns.

The proposed indicator was implemented in Java programming language and ontology was developed by a Protégé tool.[20, 21] To test the proposed ORMI, we took chart pattern of TCS share for the period of July to December 2021, i.e. for the last two quarters of the year 2021 (available at https://bit.ly/3IhFViz). A candle stick chart was used for the Bollinger band and Fibonacci retracement indicators to generate the points (entry and exit) and levels respectively, whereas the Heikin-Ashi was used for the investment or trading duration. Figure 12.7(a),(b),(c),(d) shows the candle stick chart pattern of TCS share for the period of July to December 2021.

In Figure 12.7(a), the chart was prepared without using any indicator. So, it was difficult to predict the next move of the share. Hence, no trading or investment decision could be made. Figure 12.7(b) shows the two parallel channel that are depicted by shaded region. Parallel chart is a kind of pattern based on the technical analysis. The proposed ontological indicator works when any pattern is followed by chart of any stock. Figure 12.7(c) shows the Bollinger band indicator applied on chart. This indicator indicates the entry and exit point for trading and investment. The indication triggers when band is squeezed and any candle crosses the middle line of the Bollinger band (for given the closing below or above the middle line of the Bollinger band). The former two point (shown in blue) indicate the buying opportunity for trading and investment; the latter two point (shown in pink) indicate the selling opportunity for trading and investment. Though the figure provides the trigger for entry and exit, nevertheless, it was not able to predict about some certain level for the buying and selling opportunity, and it was also not able to predict the holding period of particular trade or investment. Figure 12.7(d) shows the Fibonacci indicator applied on the chart. This indicator provides certain levels (38.2%, 50%,

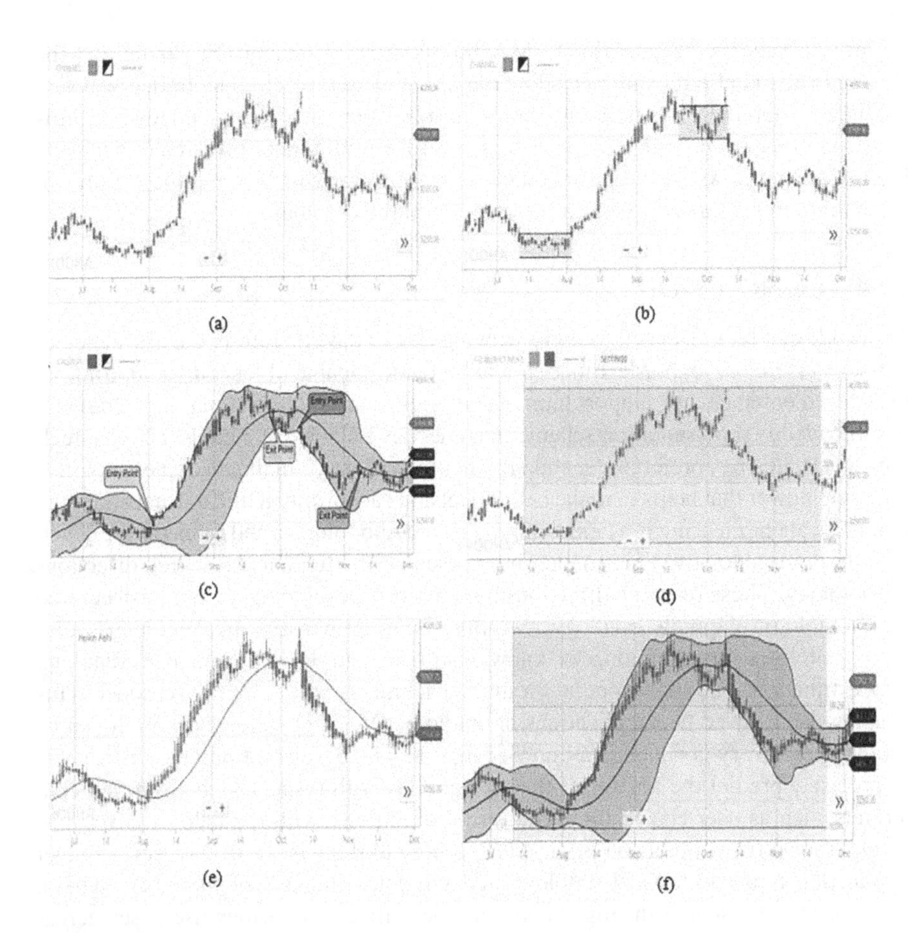

FIGURE 12.7 (a) Stock, (b) stock selection based on technical analysis, (c) Bollinger band, (d) Fibonacci retracement, (e) Heikin-Ashi, (f) ORMI.

and 61.8%) for the triggers indicating buying or selling opportunity for trading and investment decision. These levels indicated that retracement could be seen in the prince of the stock before it moves in the certain direction. With the help of these levels, traders or investors can be cautious if the price of the stock is near to these levels. Therefore, these points can be used for the profit booking or can be used as a stop loss as well as taking a new trade. However, since these points are too near and hard to predict the exact move of the stock; this was also not able to indicate the entry or exit points.

Figure 12.7(e) shows the chart containing Heikin-Ashi. The Heikin-Ashi strongly indicates the positive or negative trend of the stock that results in deciding the buying or selling opportunity to the trader and investor. This indicator triggers a buying indication if 20 EMA, the red line in Figure 12.7(e), is crossed by any green candle and the next green candle gives its closing above to the previous green candle. The selling indication is trigger if 20 EMA line is crossed by any red candle and the closing of the next red candle is below the previous red candle. The trader or investor should keep their position until the Heikin-Ashi color is changed. However, this indicator does not provide any particular level for trading and investment decision. Figure 12.7(f) shows the proposed ontological indicator which stored the features of these three indicators and pattern information in the ontology.

12.4 CONCLUSION

The work that humans need to do is growing day to day; therefore, the assistance of machines is not enough. Machines should be made capable of learning and thinking in order to enhance and support human activities in an intelligent manner. Ontology is a knowledge representation scheme that encodes knowledge, beliefs, actions, feelings, goals, desires, preferences, and all other mental states in the machine and offers inferring power that helps to make better decisions according to the current situation. Ontological engineering is a field that provides methodologies and rules for the development of an ontology. This chapter has presented the ten latest research directions for ontology. These research directions are ontology development from unstructured data, ontology mapping, ontology merging, ontology partitioning, ontology evaluation, ontology representation as knowledge unit, ontology provenance, ontology cost estimation, ontology population and enrichment, and ontology reasoner. This chapter also showed two applications of ontology. One application was developed by representing every concept as a knowledge unit. The developed ontology is capable to precisely predict the answer of the user queries with respect to the imposed constraints such as user choice for certainty vs. specificity with the help of DBA. The results showed how precision is varied according to the context of the user. Another application is used classical ontology and mitigates the risk of the stock market. The research presented in this chapter can be utilized in various use cases across multitude of domains demanding for modeling personalized behavior based on the context of the user. The reported findings are constrained by the availability of data used, which we have collected for the domain of routine tasks. Additional work is required to scale the effort for scenarios with deep-rooted hierarchy with every concept detailed to the maximum extent possible and tested for a larger number of users.

REFERENCES

1. Brachman, R. J. (1990, July). The future of knowledge representation. In *AAAI*, vol. 90, pp. 1082–1092.
2. Patel, A., & Jain, S. (2018). Formalisms of representing knowledge. *Procedia Computer Science*, 125, 542–549.
3. Gómez-Pérez, A., Fernández-López, M., & Corcho, O. (2006). *Ontological engineering: With examples from the areas of knowledge management, e-Commerce and the semantic web*. Berlin, Heidelberg: Springer Science & Business Media.
4. Antoniou, G., & Harmelen, F. V. (2004). Web ontology language: Owl. In *Handbook on ontologies*. Berlin, Heidelberg: Springer, pp. 67–92.
5. Sengupta, K., & Hitzler, P. (2014). Web ontology language (OWL). *Encyclopedia of Social Network Analysis and Mining*, 1–5.

6. Cardoso, J., & Pinto, A. M. (2015). The web ontology language (owl) and its applications. In *Encyclopedia of information science and technology* (Third Edition). United States: IGI Global, pp. 7662–7673.
7. Horrocks, I., Patel-Schneider, P. F., & Van Harmelen, F. (2003). From SHIQ and RDF to OWL: The making of a web ontology language. *Journal of Web Semantics*, 1(1), 7–26.
8. Pérez, J., Arenas, M., & Gutierrez, C. (2009). Semantics and complexity of SPARQL. *ACM Transactions on Database Systems (TODS)*, 34(3), 1–45.
9. Alterovitz, G., Xiang, M., Hill, D. P., Lomax, J., Liu, J., Cherkassky, M., . . . Ramoni, M. F. (2010). Ontology engineering. *Nature Biotechnology*, 28(2), 128–130.
10. Patel, A., Sharma, A., & Jain, S. (2019). An intelligent resource manager over terrorism knowledge base. In *Recent patents on computer science*. Netherlands: Bentham Science.
11. Michalski, R. S., & Winston, P. H. (1986). Variable precision logic. *Artificial Intelligence*, 29(2), 121–146.
12. McGrath, G. (2003). *Semantic web services delivery: Lessons from the information systems world*. In Proceedings of the 7th Pacific-Asia conference on information systems. South Australia, pp. 222–235.
13. Patel, A., Jain, S., & Shandilya, S. K. (2018). Data of semantic web as unit of knowledge. *Journal of Web Engineering*, 17(8), 647–674.
14. Jain, S., & Patel, A. (2019, October). Smart ontology-based event identification. In *2019 IEEE 13th international symposium on embedded multicore/many-core systems-on-chip (MCSoC)*. New York, NY: IEEE, pp. 135–142.
15. Otero-Cerdeira, L., Rodríguez-Martínez, F. J., & Gómez-Rodríguez, A. (2015). Ontology matching: A literature review. *Expert Systems with Applications*, 42(2), 949–971.
16. Object Properties, Link. http://protegeproject.github.io/protege/views/object-property-characteristics/
17. Data Properties, Link. https://ddooley.github.io/docs/data-properties/#
18. Bharadwaj, K. K., & Jain, N. K. (1992). Hierarchical censored production rules (HCPRs) system. *Data & Knowledge Engineering*, 8(1), 19–34.
19. Patel, A., & Jain, S. (2021). A novel approach to discover ontology alignment. *Recent Advances in Computer Science and Communications (Formerly: Recent Patents on Computer Science)*, 14(1), 273–281.
20. Tudorache, T., Noy, N. F., Tu, S., & Musen, M. A. (2008, October). Supporting collaborative ontology development in Protégé. In *International semantic web conference*. Berlin, Heidelberg: Springer, pp. 17–32.
21. Patel, A., & Jain, S. (2021). Ontology versioning framework for representing ontological concept as knowledge unit. In *ISIC*. United States: CEUR Workshop Proceedings, pp. 114–121.

13 Expert Systems in AI

Components, Applications, and Characteristics Focusing on Chatbot

Ravi Lourdusamy and Johnbenetic Gnanaprakasam

CONTENTS

DOI: 10.1201/9781003310792-13

13.1 INTRODUCTION

An expert system is a computer algorithm that can help us solve problems in the same way that humans can. In terms of decision-making speed, it may be quicker than humans. This is accomplished by querying its knowledge base using reasoning and inference techniques depending on user needs.[1] The first expert system (ES) was developed in the 1970s as part of artificial intelligence (AI). This system assists us in solving problems with a high level of complexity. All of the answers it generates are comparable to those of a knowledge expert, and this is accomplished by extracting the knowledge base that is specifically kept for this system.[1, 2] Like a human expert, the system makes conclusions about complicated issues using facts and heuristics. This system is named for its issue-solving capabilities for a certain domain, be it any level of complicated challenge in that area. This system's core design and architecture are intended for application in certain disciplines such as medicine, science, and so on.[3]

The expert system's role is to solve complicated issues and offer decision-making capabilities similar to those of a human expert. This is accomplished by the system retrieving information from its knowledge base following user queries and utilizing reasoning and inference procedures. It facilitates the distribution of human expertise. One ES might incorporate knowledge from multiple human experts, which would increase the effectiveness of the answers. It lowers the expense of seeking advice from a specialist in a variety of fields, including medical diagnosis. They employ an inference engine and knowledge base. The efficiency and functionality of the system are determined by the knowledge base that offers to it. The more information it gains, the better it performs. The finest example of an ES is the auto-correct spelling recommendations that appear in Google searches. Figure 13.1 depicts a graphical illustration of the expert system's functioning.[4]

Early research in Al (1950s–1960s) concentrated on search methods and psychological modeling. Some of that work is synthesized by expert systems, but their emphasis is now on representing and applying knowledge of particular job domains. Early research employed game-playing and reasoning about children's blocks as straightforward task domains to evaluate reasoning techniques. The focus of expert system work is on issues that have commercial or scientific significance, as

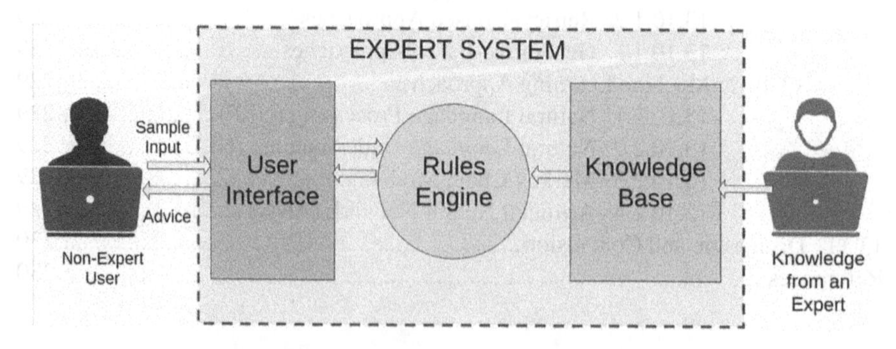

FIGURE 13.1 Expert system.

determined by people outside of AI. In the Foreword, Newell [19] refers to MYCIN as "the original expert system" since it consolidated the design considerations and placed emphasis on the application. By evaluating the benefits of current approaches and assisting in identifying their drawbacks, expert systems continue to advance and contribute to AI's research. The procedure is still in progress. Based on early work in psychological modeling, expert systems' work explored the use of production systems in the year 1975. Fundamental research on knowledge representation was transformed into practical object-oriented substrates in the 1980s.

The commercialization of expert systems has been significantly impacted by hardware advancements during the past 10 years. In tiny, cost-effective boxes, stand-alone workstations offer specialized hardware for running AI programming languages effectively, high-resolution interactive graphics, and wide address spaces. Since they do not require big, time-shared central mainframes for development and debugging, development is simpler. They also give field staff a respectable response to inquiries about portability. The final remaining obstacles to portability have effectively been removed by the development of expert systems and the languages and environments (referred to as "shells") for developing them in standard languages like Common Lisp and C.

13.2 EXPERT SYSTEM CHARACTERISTICS

As expected from any such system, the basic characteristic of an expert system is its high performance. The expert system is capable of solving any sort of complicated problem inside a specified area with high accuracy and efficiency. The response is something that defines the worth of any system, in this regard expert systems are highly understandable. The responses it provides are comprehensive and understandable to users. It can process human language input and respond to human language. A significant characteristic of an expert system is that it is reliable. The system can be trustworthy in terms of dependency and one can rely on it when it comes to producing an effective and precise result. Quick responses are another highlight of an expert system. They are highly responsive in all regards. An ES responds to any sort of complicated query in a very quick time.[5]

13.3 COMPONENTS OF AN EXPERT SYSTEM

The user interface, inference engine, and knowledge base are the three essential components of an expert system.[6] Figure 13.2 shows the components of the expert systems.

a) **User interface**
The most important component of the expert system software is the user interface. This element provides the user's query to the inference engine in a comprehensible format. In other words, an interface enables user-expert system communication. The user interface design process enables us to provide the user with clear and concise information. The system can communicate with the user, accept questions in a legible format, and send them to the

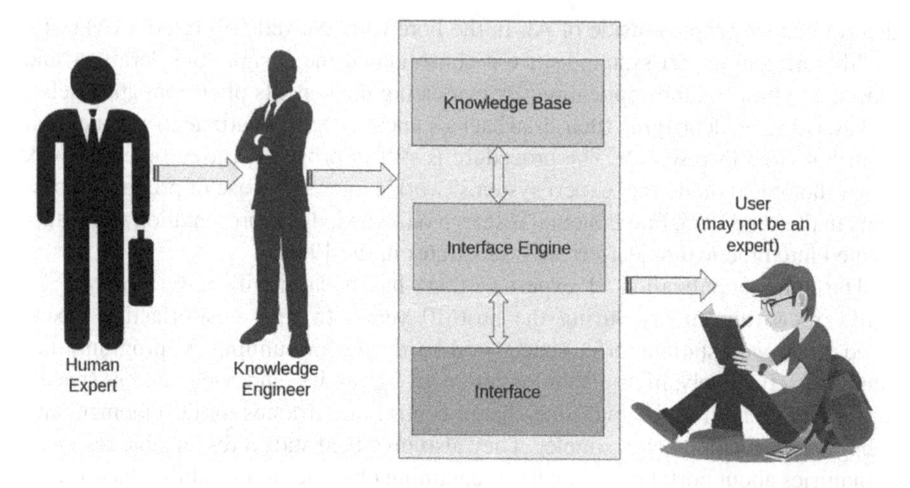

FIGURE 13.2 Components of expert system.

inference engine. After receiving a response from the inference engine, the interface provides the user's output. This means that anyone can communicate with an expert system to solve a problem.

b) **Inference engine**
The inference engine is a key part of the system, and it functions as the expert system's brain. The software uses inference rules to draw conclusions or make deductions from the knowledge base. Our software aids in the generation of correct answers for user inquiries. An inference engine is used by the system to extract knowledge from the knowledge base.

c) **Knowledge base**
Storage is used to store various types of knowledge obtained from various knowledge specialists. It is regarded as a large repository of knowledge. The more knowledge that is saved, the better the system's output. It functions similarly to a database, including all of the necessary data for every given field. For simplicity, we may think of knowledge base (KB) as a collection of objects and their properties. An eagle, for example, is an object with the characteristics of being a bird, not a tame bird, not a food source for humans, and so on.

d) **Participants in the development of the expert system**
The expert system is made up of three major participants, namely the expert, knowledge engineer, and end user. The expertise of human experts is essential to the success of an ES. These experts are people who have deep knowledge and understanding of a particular field. A knowledge engineer is someone who gathers information from domain experts and then codifies that knowledge into the system using formalism, and the end user is a specific individual or group of people who are not experts, work on the expert system, and want a solution or advice for their complex queries.[6]

13.4 CAPABILITIES OF THE EXPERT SYSTEM

Major capabilities of an expert system are discussed in this section.[4, 8]

One of the most significant features of an expert system is its capacity to advise. An expert system is so efficient that it is capable of advising humans on the query of any domain from the particular ES. Another significant feature of an expert system is its effectiveness in providing decision-making capabilities. It is capable of giving this advantage of decision-making in any domain. For instance, in situations where you need to make financial decisions or medical sciences-related decisions, expert systems can be useful. In demonstrating a device, expert systems can be of utmost help. If one needs to demonstrate features or characteristics of any new device or even to explain the specifications, expert systems can be employed. In short, they can be used as a user's manual for a product.

A significant factor of an expert system is its capability in problem-solving. In addition to the problem-solving capacity, an expert system is capable of explaining the problem to the user. It can provide a detailed description of an input problem. Interpreting the input as given by the user is another salient feature of the system, which is considered significant.

Apart from interpreting the inputs, one significant factor of expert systems is the capability to predict results, and therefore, they can be used to accurately predict results. An ES can be effectively designed for the diagnosis of a disease in the medical field. This is possible because of its inbuilt functionality. It does not need multiple components since it has various medical tools built into it, and therefore the diagnosis predicted can be accurate.

13.5 ADVANTAGES AND LIMITATIONS OF EXPERT SYSTEM

a) **Advantages of expert systems**

These systems are highly consistent and can be relied upon at any time. When it comes to practicality, these systems can be useful since they can be used in places where there is a high danger or any sort of threat to human beings. Taking the probability of errors, ES ensures that there is less chance for errors in results provided when there is a correct knowledge base given. This is possible because it is not affected by emotions, tension, or fatigue.

The availability of an expert system is an advantage. Since huge communities are working on this system, it is very easily available. For any system, no matter how effective it is, one factor that is taken into consideration is the production cost. Compared to other systems, the cost of production is reasonably cheaper and affordable. The efficiency of any system is highly accounted for by its speed and performance. Here, the processing speed is extremely fast and the amount of work is reduced. These systems are highly effective and compatible and can work in dangerous environments. These systems provide a low-risk environment for human beings, which is why they are popular.

b) **Limitations of expert system**

It is okay for any technology that provides efficient and compatible solutions to be not only expensive but also need a lot of time for development and production. It also needs a very good computing resource to function smoothly. Limitations of ES include [7, 9] difficult knowledge acquisition, maintenance costs, and development costs.

13.6 APPLICATIONS OF EXPERT SYSTEMS

One significant application of the expert system is in the manufacturing domain. It is widely applied in the construction and manufacturing of physical products like camera lenses and other automobiles. As discussed earlier, ES is a significant contribution to the medical field. It was in this field that the system was first employed. Banking and industry are other areas that are significant in contemporary society, and the financial sector ES can be effectively employed. The fraud detection tool is used in the banking industry to identify any fraudulent activity and to provide banking advice on whether or not to offer loans to a company. To achieve a task's aim, expert systems can be used. It can be effectively used to plan and schedule activities. In this regard, one significant application of ES can be in planning and scheduling.

Agriculture, education, the environment, law, industry, medical, and power systems are just a few of the key application areas for expert systems. An expert system is used to create a huge number of new products as well as new configurations of existing ones.[5, 7, 8, 10] Different types of applications and their roles are tabulated in Table 13.1.

The field of medicine adopted one of the first expert system applications. The developed expert system, called MYCIN, was designed to recognize numerous bacteria that can result in life-threatening illnesses and to make medication recommendations depending on the patient's age, sex, and weight. Backward chaining, often known as a goal-directed control approach, is predominantly used by MYCIN. The argument's logical validity is demonstrated in the same manner, yet the system behaves very differently. In goal-directed reasoning, a system begins with a description of the desired

TABLE 13.1
Expert System Applications and Their Roles

Applications	Role
Design Domain	Lens design for cameras vehicle design
Medical Domain	Human surgical procedures are performed as part of disease diagnosis systems that make use of visible data to detect the cause of the illness
Monitoring Systems	Comparing data to observable systems continuously
Process Control Systems	Control of physical processes based on monitoring
Knowledge Domain	Finding errors in computers or autos
Commerce	Identifying possible fraud and suspicious transactions, investing in the stock market, airline scheduling and cargo scheduling

outcome and proceeds "backward," or from right to left, through the inference rules to identify the facts necessary to reach that outcome.

Simplified Logical "Hierarchy" of MYCIN
Find out about C
If B, then C
If A, then B
Is A true?

This is an illustration of how knowledge is primarily represented by conditional statements or rules in the algorithm (simplest version) of MYCIN. In this case, learning more about "C," which might be a specific illness, is the objective. According to Rule 1, if "B" is true, "C" must likewise be true. B in this case can be viewed as a symptom. Therefore, MYCIN is searching for a specific symptom in a patient called "B" that can be linked to his or her illness and identifies the condition as "C." Additionally, it has an implicit rule that states that "A causes B," meaning that if the patient engaged in a particular behavior, "A," then it is possible to anticipate the symptom, "B." Consider the following fictitious example: MYCIN is treating a patient who is infected with the HIV, which causes AIDS. It will check for typical AIDS symptoms like fatigue or appetite loss. It will next ask the patient if he or she has engaged in any unprotected sex or other similar recognized AIDS risks to support its argument. If the response is "yes," MYCIN will conclude that the patient has AIDS using its rule-based approach. Remember that this is an extremely basic illustration of how an expert system like MYCIN might operate. In practice, it may be necessary to take into account hundreds of variables, build a large number of rules or algorithms as a result, and then analyze a vast amount of data to identify symptoms and diseases.

The talent acquisition/human resource sector just developed another breakthrough of an expert system. Wade & Wendy is an AI-based expert system that is trained to speak with a client (company) to gather the precise data required to find a suitable applicant to fill the organization's hiring needs. Wade & Wendy's heuristics are driven by knowledge-based algorithms. Wade & Wendy will continue to develop and learn as it engages in more talks, much like a human recruiter strives to have meaningful interactions with CEOs and managers in order to completely understand and adapt to their changing needs. This specific expert system is an illustration of a particular class of AI known as a chatbot: a computer program or artificial intelligence that engages in conversation via oral or written means. These algorithms frequently pass the Turing test because they accurately mimic how a human would act as a conversational partner. Chatbots are frequently employed in dialogue systems for a variety of useful tasks, such as information gathering or customer assistance. Here, the fields of NLP and ES and artificial intelligence converge.

We are moving closer to a time when AI and people will collaborate to solve issues. To make expert systems palatable to the general public, however, a number of difficult problems still need to be resolved. Expert systems, a subfield of AI, are primarily in charge of assisting and learning from human experts. Finding a balance between the work of human experts in many sectors and the capabilities of AI will need a lot of effort and research. Although the usage of expert systems in the industry is still uncommon, it has significantly increased in recent years with the rise of big data as more and more businesses test the potential of AI.

13.7 TRADITIONAL SYSTEMS VERSUS EXPERT SYSTEMS

The way problem-related expertise is encoded differs significantly between standard and expert systems. Issuing expertise is commonplace in conventional applications, with both codes and data structures featuring prominently. The problem-related expertise method is a well-known technique used in expert systems. Furthermore, applying knowledge in expert systems is crucial. Expert systems are better equipped to use data than traditional systems.[5]

The incapacity of traditional methods to offer reasons for issue resolution is one of their most significant drawbacks. This is because these systems try to meet the challenge in a simple way. Expert systems can provide explanations and make understanding a particular response simpler than humans can.[9]

An expert system, in general, does computations using symbolic representations. On the contrary, traditional systems cannot express these terms. They can only provide simplified solutions without answering the "why" and "how" questions. Furthermore, the system's specialists are supplied with problem-solving tools that are more intelligent than traditional ones, meaning they are better equipped to address various challenges.

Compared to a conventional system that relies on human specialists, an expert system offers many benefits. Even though expert systems rely on knowledge from human specialists, human intelligence is brittle. For instance, if the human expert is an employee who needs to take a leave of absence, people who need to address a problem no longer have access to that source of information. Due to its reliance on a persistent repository of knowledge that can be accessed repeatedly, expert systems are more reliable and independent of the user's level of expertise. In the same scenario, any agent or worker can utilize the expert system to quickly resolve a problem even if a human expert is not there.

The nature of human specialist knowledge is likewise unexpected. The human specialist may operate at a lower level than usual on days when they aren't feeling well. An expert system, however, consistently delivers at a high level regardless of who utilizes it, providing consistent performance.

Additionally, it is challenging to replicate human knowledge through various forms of communication. Take the example of a human specialist who is required to help with a sizable number of calls that require technical support. Only one consumer at a time can be assisted by a human over a single communication channel. As opposed to an automated expert system that can intelligently apply knowledge across all channels and support numerous callers simultaneously, this one is less capable of doing so. An expert system is a crucial tool for businesses that deal with a significant volume of customer support calls to ensure correct and effective handling of those calls.

Documenting and distributing human knowledge systems across a complicated network is challenging. Even when transmitted properly, the knowledge might not be conveyed to non-experts at the intended degree of expertise or at all, depending on the circumstances. The knowledge base of an expert system, on the other hand, is straightforward to deploy over the entirety of a system of many channels and is easily documented for non-experts.

Last, but not least, maintaining human specialist resources is expensive. Experts demand higher wages, therefore businesses may need to bring on more than one to have enough problem-solving capabilities on hand. As long as the system is functioning properly, an expert system is far more cost-effective because it can store and automatically share the knowledge of a human expert. Once the expert system is operational, organizations won't need to rely on the expensive labor of human experts to maintain a smooth operation.

Knowledge categories expert systems take into account a wide range of knowledge types. Several of these combine to generate dimensions of divergent knowledge that is both in-depth and meta-knowledge: domain knowledge; common sense or general knowledge; implicit knowledge; heuristics; algorithms; knowledge that is procedural, declarative or semantic; and public or private information that is superficial information.

13.8 STEPS TO DEVELOP AN EXPERT SYSTEM (ES)

Six steps are necessary for the development of an expert system, as shown in Figure 13.3.

In the first step, the problem must be identified, and it should be suitable for an expert system to give a solution. ES experts of the domain need to be identified and the cost-effectiveness needs to be identified at this stage. Identifying ES technology is the second step. From the experts of ES, acquiring the domain-specific knowledge for developing the system is done in the third step. Testing and refining the prototype of the ES is the fourth step. The fifth step is developing, documenting, and giving training for the users of ES. Maintaining the ES system and keeping the knowledge base up-to-date is the final step of an ES.

13.9 A DIFFERENT PERSPECTIVE ON EXPERT SYSTEM ARCHITECTURE

The key components of a rule-based expert system are depicted as shown in Figure 13.1. Through a user interface, the user communicates with the system (which may employ menus, spoken language, or any other kind of communication). Then, using both the expert knowledge (derived from our friendly expert) and data related to the particular problem being solved, an inference engine is employed to reason. A series of IF-THEN rules will typically represent the expert knowledge. The user-provided data and any partial inferences (along with certainty measures) drawn from it are both included in the case-specific data. The case-specific information will be the components in working memory in a straightforward forward chaining rule-based system. Nearly all expert systems also include an explanation subsystem that enables the software to communicate its logic to the user. A knowledge base editor is another feature that some systems offer, making it easier for experts and knowledge engineers to update and review the knowledge base.

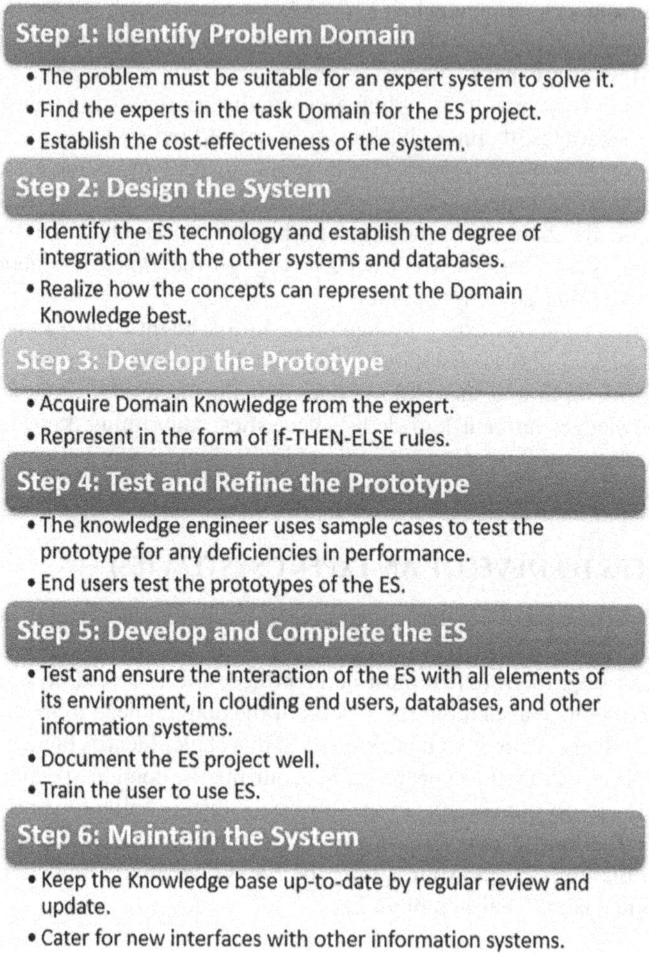

Step 1: Identify Problem Domain

- The problem must be suitable for an expert system to solve it.
- Find the experts in the task Domain for the ES project.
- Establish the cost-effectiveness of the system.

Step 2: Design the System

- Identify the ES technology and establish the degree of integration with the other systems and databases.
- Realize how the concepts can represent the Domain Knowledge best.

Step 3: Develop the Prototype

- Acquire Domain Knowledge from the expert.
- Represent in the form of If-THEN-ELSE rules.

Step 4: Test and Refine the Prototype

- The knowledge engineer uses sample cases to test the prototype for any deficiencies in performance.
- End users test the prototypes of the ES.

Step 5: Develop and Complete the ES

- Test and ensure the interaction of the ES with all elements of its environment, in clouding end users, databases, and other information systems.
- Document the ES project well.
- Train the user to use ES.

Step 6: Maintain the System

- Keep the Knowledge base up-to-date by regular review and update.
- Cater for new interfaces with other information systems.

FIGURE 13.3 Expert system designing steps.

Expert systems' ability to distinguish between more general-purpose reasoning and representation techniques and domain-specific knowledge is a key feature. An expert system shell is a term used to describe the general-purpose bit (in the dotted box as shown in Figure 13.1). The inference engine (and knowledge representation method), a user interface, an explanation system, and occasionally a knowledge base editor are all provided by the shell, as shown in Figure 13.1. When faced with a novel challenge (like automobile design), we can frequently identify a shell that offers the appropriate level of support; all that is left to do is contribute the necessary specialist knowledge. There are many commercial expert system shells, and they are each suitable for a distinct set of issues. Developing expert system shells and writing expert systems that use shells are both parts of the industry's expert systems work.

13.9.1 Typical Expert System Tasks

The types of problems that can be solved by an expert system are not fundamentally constrained. Typical activities performed by expert systems today include data interpretation, including the use of geophysical readings or sonar data; identification of defects, including those in machinery or human disorders; the structural analysis or arrangement of complicated structures, such as computer systems or chemical compounds; creating action plans that could be carried out by robots; and making future predictions about the weather, stock prices, and exchange rates. However, "traditional" computer systems today are also capable of performing some of these tasks.

13.9.2 Problems with Expert Systems

Human specialists possess common sense in addition to a wealth of technical expertise. How to offer expert systems common sense is still a mystery. Expert systems are incapable of displaying creativity in response to novel circumstances, while human experts can.

13.9.2.1 Learning

Expert systems must intentionally update, whereas human experts adapt to new settings spontaneously. Neural networks and case-based reasoning are two techniques that can combine learning.

13.9.2.2 Sensory Experience

Expert systems currently rely on symbolic input, whereas human experts can access a wide range of sensory experiences.

13.9.2.3 Degradation

Expert systems struggle to identify situations where there is no solution or when their expertise is not needed.

As can be seen, developing expert systems is a complex process that takes into account a variety of elements. The inclusion of people (human experts), who have little comprehension of AI and struggle to communicate their specific expertise to AI engineers, is one of the main obstacles to making expert systems work. On the other hand, AI developers could struggle to fully comprehend a human expert's cognitive process in a particular field, like engineering. To implement expert systems, a company would need to persuade a legitimate human expert to work with the AI development team. This could occasionally turn out to be a challenging circumstance. The following issues are listed in the article "Why Expert Systems Fail" from *The Journal of the Operational Research Society*.[20] The human specialist might not be accessible. An expert may need to spend a lot of time learning new things and taking him or her away from his or her usual tasks may be too expensive for the firm. The human expert is reluctant to share his or her ideas. The expert could feel frightened if he or she perceives AI development as a replacement.

The same article describes how expert systems lack a human expert's "common sense" For instance, a medical sector expert system might not have a specific condition in its knowledge base, and it might mistakenly label a patient as healthy even though they are in pain and appear to be unwell.

Expert systems' stringent domain dependence is another issue, claims an essay titled "Diagnostic Expert Systems: From Expert's Knowledge to Real-Time Systems". [21] This can be challenging when the expert system is attempting to address an issue that may require knowledge that is not within its purview.

Additionally, in fields like engineering and medicine, it may take knowledge from several different branches to solve a single problem. According to the study "Expert Systems: Perils and Promise" expert knowledge is occasionally shared.[22] Reliance on any one expert can either lead to the creation of blind spots in the knowledge base or to a system that will not have users, according to experience with systems that have made it through the feasibility demonstration stage. Communities of experts that share their knowledge to solve problems frequently exist.

Expert systems have existed since the early 1980s, but until recently, they were simply not reliable enough. This is mostly due to the fact that expert systems need a large amount of data to function, which is now easily accessible thanks to the recent big data revolution led by powerful Internet companies like Google, Amazon, Yahoo, and Facebook, among others. This fixes the data problem and makes it possible for businesses like Google and Amazon to function. However, many of the most serious issues with expert systems that were raised in this section have not yet been resolved and are still being investigated.

13.10 CHATBOT DEVELOPMENT APPROACHES/CHATBOT DESIGN TECHNIQUES

Chatbots are classified into three types: simple chatbots, intelligent chatbots, and hybrid chatbots. Simple chatbots with limited features are sometimes referred to as rule-based bots. They are task-oriented.[11] AI-powered intelligent chatbots are intended to mimic human-like interactions. Hybrid chatbots are a mix of basic and intelligent chatbots.

Developing a chatbot depends on the algorithm, which is used to create a chatbot, and the technique implemented, which the chatbot developer uses. The chatbot landscape today is vast. Chatbots are not classified as a specific category, but rather as part of a larger range. The following taxonomy is proposed based on input or message routes. Different factors may be used to classify chatbots.[12]

13.10.1 Chatbot Development Approaches

There are several different ways a person can create a task-oriented or non-task-oriented chatbot. Both chatbots can use these tactics interchangeably.

13.10.1.1 Rule-Based Approaches

Decision-tree bots are a type of chatbot that use rules to make decisions. Their adherence to a set of guidelines guarantees precision and accuracy. These guidelines will help the person address common issues that the bot is very familiar with. Rule-based chatbots plan conversations in a way that is similar to how flowcharts work.[13, 14]

13.10.1.2 Retrieval-Based Approaches

Recovery-based chatbots use techniques such as keyword matching, machine learning, or deep learning to determine the most appropriate answer. Regardless of the approach, these chatbots only provide predefined answers and do not produce new results. Mitsuku is a great example of a retrieval-based chatbot that can help you get the information you need quickly.[15]

13.10.1.3 Generative-Based Approaches

Predefined replies are the only possibility for retrieval-based systems. Generic chatbots may produce new conversations based on huge amounts of conversational training examples. Generic chatbots use a variety of methods to learn, including supervised learning, unsupervised learning, reinforcement learning, and adversarial learning.[16]

Parsing, pattern matching, rule-based chatbots/pattern matching are ways to implement chatbots. The three most common languages, namely AIML, RiveScript and ChatScript, and ontologies are the techniques used to build chatbots.

13.10.2 Machine Learning Approach

13.10.2.1 Natural Language Processing (NLP)

Natural language processing (NLP) is the process of synthesizing and analyzing human languages. NLP helps chatbots learn how to communicate in a way that is similar to how humans communicate. It creates the impression that you are conversing with a person rather than a machine.[17]

13.10.2.2 Natural Language Understanding (NLU)

Understanding the meaning of the user's input is what NLU is all about. NLU primarily deals with machine reading comprehension and allows chatbots to understand the meaning of the body of the text. NLU is nothing more than comprehending the provided text and categorizing it into appropriate intents. NLU may be used in a variety of procedures.[18]

13.10.2.3 Markov Chain Model

It is a fundamental idea that may be used to explain even the most complicated real-time events. This basic technique known as the Markov chain is employed in some form or another by chatbots, text IDs, text creation, and many more AI algorithms.[12]

13.10.2.4 Artificial Neural Network (ANN)

The following are the subsystems of ANN: gated recurrent unit, recurrent neural network (RNN), Seq2seq, DeepSqe2seq, and long-short term memory (LSTM).

13.11 DISCUSSION AND CONCLUSION

An ES is not going to replace human beings in any way; it is just going to improve the accuracy and the time factor that is required to make a decision. It requires a human expert to feed it with a knowledge base, and it is going to make the process quicker. The technologies and approaches used to design an AI-powered intelligent chatbot system were explained in this chapter. In the current education system, educators find it difficult to use technology for the effective teaching-learning and evaluation process. Chatbots plays a wide role compared with the other techniques or platforms. It also creates interest among the students to learn by themselves. This chapter provides a space for the readers and the developers to understand the basics of expert system.

REFERENCES

1. Akma, N., et al. (2018). Review of chatbots design techniques. *International Journal of Computer Applications*, 181(8), 7–10.
2. Drăgulescu, D., & Albu, A. (2007). Expert system for medical predictions. In *SACI 2007: 4th international symposium on applied computational intelligence and informatics—proceedings*, pp. 123–128.
3. Duda, R. O., & Shortliffe, E. H. (1983). Expert systems research. *Science*, 220(4594), 261–268.
4. Eisman, E. M., Navarro, M., & Castro, J. L. (2016). A multi-agent conversational system with heterogeneous data sources access. *Expert Systems with Applications*, 53, 172–191.
5. Grif, M., & Ayush, Y. (2017). Data analysis of expert systems by pulse diagnosis. In *Proceedings—2016 11th international forum on strategic technology, IFOST 2016*, vol. 1, pp. 329–332.
6. Fiksel, J., & Hayes-Roth, F. (1989). Knowledge systems for planning support. *IEEE Expert*, 4(3), 16–23, doi: 10.1109/64.43267.
7. Kaisler, S. H. (1986). Expert systems: An overview. *IEEE Journal of Oceanic Engineering*, 11(4), 442–448.
8. Li, Y., Huang, C., & Lu, C. (2020). Research on expert system in power network operation ticket. In *2020 IEEE international conference on artificial intelligence and computer applications (ICAICA)*. Dalian, China: IEEE, pp. 1091–1095. doi: 10.1109/ICAICA50127.2020.9182633.
9. Marshall, J. A. (1993). *Artificial intelligence/expert systems: A teaching tool.* In: Proceedings of IEEE Frontiers in Education Conference-FIE'93.
10. McShane, M. (2017). Natural language understanding (NLU, Not NLP) in cognitive systems. *AI Magazine*, 38(4), 43–56.
11. Molnár, G., & Szüts, Z. (2018). The role of chatbots in formal education. In *2018 IEEE 16th international symposium on intelligent systems and informatics (SISY)*. Subotica, Serbia: IEEE, pp. 000197–000202. doi: 10.1109/SISY.2018.8524609.
12. Nimavat, K., & Champaneria, T. (2017). Chatbots: An overview types, architecture, tools and future possibilities. *International Journal of Scientific Research and Development*, 5(7), 1019–1026.
13. Noguchi, T., et al. (2018). A practical use of expert system "AI-Q" focused on creating training data. In *2018 5th international conference on business and industrial research (ICBIR)*. IEEE Conference Location: Bangkok, Thailand, pp. 73–76. doi: 10.1109/ICBIR.2018.8391169.
14. Gbenga, O., Okedigba, T., & Oluwatobi, H. (2020). An improved rapid response model for university admission enquiry system using chatbot. *International Journal of Computer (IJC)*, 38(1), 123–131. www.researchgate.net/publication/342248071_An_Improved_Rapid_Response_Model_for_University_Admission_Enquiry_System_Using_Chatbot.

15. Okonkwo, C. W., & Ade-Ibijola, A. (2021). Chatbots applications in education: A systematic review. *Computers and Education: Artificial Intelligence*, 2, 100033. https://doi.org/10.1016/j.caeai.2021.100033
16. Pandit, V. B. (1994). Artificial intelligence and expert systems: A technology update. In *Conference proceedings 10th anniversary IMTC/9, advanced technologies in I & M. 1994 IEEE instrumentation and measurement technolgy conference (Cat. No.94CH3424-9)*. IEEE Conference Location: Hamamatsu, Japan, vol. 1, pp. 77–81. doi: 10.1109/IMTC.1994.352122.
17. Patel, Y. S., Vyas, S., & Dwivedi, A. K. (2015). A expert system based novel framework to detect and solve the problems in home appliances by using wireless sensors. In *2015 international conference on futuristic trends on computational analysis and knowledge management (ABLAZE)*. IEEE Conference Location: Greater Noida, India, pp. 459–464. doi: 10.1109/ABLAZE.2015.7155039.
18. Song, Y., et al. (2018, July). An ensemble of retrieval-based and generation-based human-computer conversation systems. *IJCAI International Joint Conference on Artificial Intelligence*, 4382–4388.
19. Buchanan, B. G., & Shortliffe, E. H. (1984). Rule-based Expert Systems: The MYCIN Experiments of the Stanford Heuristic Programming Project, Addison-Wesley series in artificial intelligence. Addison-Wesley Longman Publishing Co., Inc. ISBN: 0201101726, 9780201101720.
20. Michael Z. B. (1985). Why expert systems ail. *The Journal of the Operational Research Society*, 36(7), 613–619. https://doi.org/10.2307/2582480
21. Angeli, C. (2010). Diagnostic expert systems: From expert's knowledge to real-time systems. *Advanced Knowledge Based Systems: Model, Applications & Research*, 1, pp. 50–73.
22. Bobrow, D. G., Mittal, S., & Stefik, M. J. (1986). Expert systems: Perils and promise. *Communications ACM*, 29(9), 880–894. https://doi.org/10.1145/6592.6597.

Index

K

L

M

N

O

P

Q

R

S

T

U

V

W